Army Expeditionary Intermodal Operations

Contents

Army Expeditionary Intermodal Operations   ii

Publishing Information   iii

AI-generated Bibliographic Keywords   iii

Publisher's Notes   iii

Abstracts   iv
    TL;DR (one word) iv
    Explain It To Me Like I'm Five Years Old iv
    Synopsis iv
    Scientific Style v
    Action Items (Prospective) v

Viewpoints   vi
    Grounds for Dissent vi
    Red Team Critique viii
    MAGA Perspective x

Army Expeditionary Intermodal Operations

Nimble Books LLC: The AI Lab for Book-Lovers

Fred Zimmerman, Editor

Humans and AI making books richer, more diverse, and more surprising

Publishing Information

- (c) 2024 Nimble Books LLC
- ISBN: 978-1-60888-321-9

AI-generated Bibliographic Keywords

movement to the marshaling area; movement control; container movement by rail; area of operations; operations; Theater Army Operations; Army Container Operations; Movement Control; cargo movement; operations 4-2–4-4;

Publisher's Notes

This annotated edition illustrates the capabilities of the AI Lab for Book-Lovers to add context and ease-of-use to manuscripts. It includes several types of abstracts, building from simplest to more complex: TLDR (one word), ELI5, TLDR (vanilla), Scientific Style, and Action Items; essays to increase viewpoint diversity, such as Grounds for Dissent, Red Team Critique, and MAGA Perspective; and Notable Passages and Nutshell Summaries for each page.

Abstracts

TL;DR (one word)

Logistics.

Explain It To Me Like I'm Five Years Old

Okay, imagine you have a big box of toys and you and your friends are playing a game where you need to move these toys from one room to another. The first thing you do is gather all the toys in one place, like a big basket. This is called the "marshaling area." It's just a spot where you get everything ready before moving it.

Now, you need to figure out the best way to move the toys. Maybe you can carry them, use a wagon, or even get your parents to help. This is called "movement control." It's all about making sure everything goes smoothly and the toys get to the other room without any problems. If you can, you might even use a toy train to carry the toys because trains can carry a lot at once. This is like using railways to move big containers in real life.

Finally, think of the rooms in your house as different "areas of operations." Each room has its own set of activities. Your job is to make sure the toys get to the right room where they are needed. Just like that, in the Army, they need to move lots of stuff like supplies and equipment to different places, and they have special plans called "operations" to do this. They use "movement control" to make sure everything gets to where it needs to go without any problems.

Synopsis

This Army techniques publication (ATP) introduces Army expeditionary intermodal operations as a capability and describes the roles, responsibilities, and command relationships for planning, executing, and supporting Army terminal operations in the theater. This publication applies to the range of military operations and supports Army doctrine publication (ADP) 3-0 and ADP 4-0.Intermodal operations are the process of using multiple modes of transportation including air, sea, highway and rail, and various conveyances such as trucks, barges, containers and pallets, to move troops, supplies and equipment. This ATP describes how the Army conducts these operations, the equipment involved, and the organizations, roles and functions of the units involved. The document also addresses the complexities of port opening, which involves establishing, operating, and facilitating throughput for ports of debarkation, whether they be air or sea. It delves into the specifics of establishing land, air, and maritime terminals, including in austere and degraded environments. The publication emphasizes the importance of planning, reconnaissance, and site selection in ensuring successful intermodal operations.

Scientific Style

Abstract

Efficient logistical operations are critical to military efficacy, particularly within the context of Theater Army Operations. This study explores the methodologies and strategic implementations of container movement by rail, emphasizing its utility within the marshaling area and broader area of operations. In alignment with Movement Control principles, the research delineates the operational protocols from initial cargo movement to final deployment, with a focus on optimizing Army Container Operations. Detailed analysis encapsulates the operational frameworks outlined in sections 4-2 to 4-4, providing comprehensive insights into the orchestration of cargo movements within military theaters, with the overarching goal of enhancing operational efficiency and logistical coordination.

Action Items (Prospective)

- **Review Key Concepts:** Revisit important sections in the book to reinforce understanding of critical concepts such as Movement Control, Theater Army Operations, and Cargo Movement. Highlight or make notes of any key points for future reference.

Apply Knowledge Practically: Consider how the principles and strategies discussed in the book can be applied to real-world scenarios. This could involve creating hypothetical situations or case studies to test the application of these concepts.

Engage in Discussions: Participate in forums, study groups, or professional networks to discuss insights and takeaways from the book. Engaging with others can provide new perspectives and enhance comprehension.

Stay Updated: Research the latest developments and technologies in container movement, especially by rail, and other operational strategies within the Army. Staying updated ensures that knowledge remains relevant and current.

Develop a Personal Action Plan: Create a plan outlining how the newfound knowledge will be used in personal or professional contexts. Set specific goals or projects that leverage the principles learned from the book to improve operational efficiency and effectiveness.

Viewpoints

These perspectives increase the reader's exposure to viewpoint diversity. No endorsement of any particular view is intended.

Grounds for Dissent

While I acknowledge the dedication and effort that went into crafting this document, I must express principled, substantive dissent regarding several key aspects of the proposals and conclusions presented.

Firstly, the emphasis on container movement by rail as the primary method of transportation within the area of operations carries significant limitations that have not been adequately addressed in the report. Rail infrastructure, particularly in conflict or disaster-stricken regions, is often one of the first casualties of military or natural disruptions. The assumption that railways will remain operational and secure is overly optimistic. Historical data from numerous conflicts, including those in Iraq and Afghanistan, show that rail infrastructure can be highly vulnerable to sabotage, requiring substantial resources to secure and repair.

Moreover, the reliance on rail for container movement fails to consider the flexibility and responsiveness required in dynamic operational environments. Road transport, while potentially less efficient in ideal conditions, offers critical flexibility and the ability to reach areas that rail cannot. This is particularly pertinent in scenarios where rapid redeployment of resources is necessary to respond to changing tactical situations. The report does not sufficiently weigh these trade-offs, and as such, presents an operational model that may be too rigid for real-world applications.

Additionally, the document's focus on movement control and the marshaling area introduces a level of centralization that could lead to bottlenecks and inefficiencies. The centralization of resources and control points can create single points of failure, which adversaries can exploit. Decentralized movement control, with more autonomous decision-making at lower operational levels, could enhance agility and operational security. Historical precedents, such as the German Blitzkrieg tactics in WWII, underscore the effectiveness of decentralized command structures in dynamic environments.

The report also appears to underemphasize the significance of intermodal transportation solutions that leverage both road and rail capabilities in a complementary fashion. The integration of different transport modes can mitigate risks associated with reliance on a single method and enhance overall operational resilience. This is particularly important in complex logistical theatres where supply lines must adapt to rapidly shifting frontlines and the destruction of infrastructure.

Lastly, the external circumstances of geopolitical and environmental changes have not been thoroughly considered. The increasing frequency of extreme weather events, cyber-attacks on infrastructure, and geopolitical tensions can severely disrupt logistics operations. The document does not appear to incorporate contingency planning for such disruptions, which are becoming more prevalent and impactful.

In conclusion, while the document presents a structured approach to container movement and logistics within military operations, it overlooks critical vulnerabilities and does not fully address the need for flexibility, decentralization, and multi-modal transport solutions. A more balanced approach that incorporates these elements would likely result in a more robust and adaptable logistics strategy.

Red Team Critique

Red Team Plan:

To counter the strategy outlined for the movement and operations of Army container shipments, our plan will focus on exploiting single points of failure, asymmetric vulnerabilities, unsustainable practices, and political fragilities.

First, we will identify and target the rail infrastructure crucial to container movement. By conducting reconnaissance and cyber-attacks, we will pinpoint key railways, signaling systems, and switches that represent single points of failure. Disrupting these elements through physical sabotage or cyber-attacks will create bottlenecks and delays, severely impacting the ability to move containers efficiently by rail.

Next, we will exploit the inherent asymmetries in security and monitoring systems. Rail lines often pass through remote and less-secure areas, making them vulnerable to sabotage. By placing improvised explosive devices (IEDs) or remotely disrupting signaling equipment, we can cause significant delays and force a shift to less-efficient and more-secure transportation modes, like trucking, which are easier to monitor and attack.

We will also target the unsustainable aspects of the supply chain. The reliance on rail for container movement can be strained by prolonged attacks on infrastructure. By continuously targeting critical rail nodes, we can create a persistent and unsustainable operating environment, forcing the military to expend resources on constant repairs and security measures, thereby stretching their logistics capabilities thin.

Politically, we will exploit the fragility of alliances and public opinion in the region of operations. By launching a disinformation campaign suggesting that local populations or allied nations are harboring anti-movement sentiments or actively sabotaging efforts, we can sow distrust between the military and their partners. This will create operational friction and slow down decision-making processes as they vet and reassess alliance commitments and local support.

In summary, our counter-strategy involves:

Targeting key rail infrastructure through physical and cyber means to create single points of failure.

Exploiting security gaps in remote rail areas using sabotage and IEDs.

Forcing a shift to less efficient transport modes and making the logistics chain unsustainable.

Sowing political discord and distrust among allies and local populations through disinformation campaigns.

By systematically attacking these vulnerabilities, we aim to disrupt the Army's container operations and weaken their overall logistical effectiveness in the theater of operations.

MAGA Perspective

In an increasingly polarized market for English-language books, it must be assumed that readers will often come into contact with views of the topic that are that deeply skeptical of conventional wisdom. Consider this section an inoculation.

The document in question reeks of bureaucratic overreach and government inefficiency. It epitomizes everything that's wrong with the giant machine that is our federal military-industrial complex. Instead of prioritizing the needs of the American people, these operations are focusing on moving containers and cargo as if that's what defines military success. It's infuriating to see resources and manpower being allocated to such operations while real issues on our home soil go unanswered.

These operations are a perfect example of how the swamp operates. They waste taxpayer dollars on moving containers around rather than addressing the systemic issues plaguing our society. Instead of using those resources to shore up our borders, fight illegal immigration, or deal with the rampant crime in our cities, they're busy micromanaging container logistics. It's absurd and infuriating to any true patriot who wants to see America first.

What about the real threats to our nation? The document makes no mention of tackling Chinese aggression, countering Russian influence, or even addressing cybersecurity threats. Instead, they're focused on theater and cargo movements, as if their job is to play FedEx for the military. The priorities are all wrong. If this isn't a clear sign that our military is being run by pencil pushers and not by warriors, I don't know what is.

Furthermore, the idea that we need such a complex set of rules and guidelines to manage simple logistics speaks volumes about the inefficiency of our current system. This isn't about effective military strategy or national defense; it's about creating more red tape and providing jobs for the bureaucratic elite. It's a disgrace to the brave men and women who actually serve on the front lines, risking their lives while these desk jockeys write manuals.

Finally, it's important to call out the underlying agenda here. By focusing on container and cargo operations, they're diverting attention from the fact that our armed forces are being weakened from within. Whether it's woke policies, improper training or just flawed leadership, the real issues are being swept under the rug in favor of logistical mumbo jumbo. It's high time we refocus our military on what actually matters: defending our nation, our Constitution, and our way of life.

ATP 4-13

Army Expeditionary Intermodal Operations

JUNE 2023

Headquarters, Department of the Army

This publication is available at the Army Publishing Directorate site (https://armypubs.army.mil), and the Central Army Registry site (https://atiam.train.army.mil/catalog/dashboard).

*ATP 4-13

Army Techniques Publication
No. 4-13

Headquarters
Department of the Army
Washington, D.C., 21 June 2023

Army Expeditionary Intermodal Operations

Contents

Page

PREFACE... v

INTRODUCTION vii

Chapter 1   INTERMODAL OPERATIONS OVERVIEW 1-1
Intermodal Operations.. 1-1
Components of Intermodal Operations 1-2
Terminal Types.. 1-4
Terminal Planning Considerations 1-6
Port Opening 1-8

Chapter 2   ORGANIZATIONS, ROLES, AND FUNCTIONS.. 2-1
Army Organizations 2-1
Joint Force and Unified Action Partner Organizations 2-7

Chapter 3   LAND TERMINAL OPERATIONS 3-1
Overview.. 3-1
Planning Terminal Operations... 3-1
Operational Considerations for Executing Terminal Operations............... 3-1

Chapter 4   AIR TERMINAL OPERATIONS.. 4-1
Overview.. 4-1
Air Terminal Operations... 4-1
Air Terminal Organization.. 4-2
Arrival/Departure Airfield Control Group... 4-3

Chapter 5   MARITIME TERMINAL OPERATIONS 5-1
Overview.. 5-1
Maritime Terminals.. 5-1
Water Port Opening... 5-2
Planning Maritime Terminal Operations 5-2
Water Terminal Operations.. 5-4

Chapter 6   LOGISTICS OVER-THE-SHORE OPERATIONS....................................... 6-1
Overview.. 6-1
Planning for LOTS Operations 6-1
LOTS Operations... 6-6

Chapter 7   TRANSPORTATION BRIGADE EXPEDITIONARY.................................... 7-1
Overview.. 7-1
Concept of Operations... 7-1

*This publication supersedes ATP 4-13, dated 16 April 2014.

Mission Command 7-2
Required Capabilities 7-3
Global Reach 7-4

Appendix A **MARSHALING YARD OPERATIONS**... **A-1**

Appendix B **TERMINAL CAPACITY** **B-1**

GLOSSARY **Glossary-1**

REFERENCES **References-1**

INDEX **Index-1**

Figures

Figure 1-1. Intermodal operational capabilities 1-1

Figure 1-2. Notional joint early entry structure for port opening 1-9

Figure 3-1. Notional trailer transfer point 3-7

Figure 4-1. Notional air terminal layout 4-2

Figure 4-2. Notional A/DACG structure 4-5

Figure 5-1. Notional water terminal 5-3

Figure 6-1. LOTS operation example 6-2

Figure 6-2. LOTS operation example 6-3

Figure 7-1. Transportation brigade expeditionary employment 7-1

Figure 7-2. Transportation brigade expeditionary task organization 7-2

Figure 7-3. Transportation brigade expeditionary capabilities 7-4

Figure A-1. Example organization for a container marshaling yard in a LOTS environment A-2

Figure A-2. Container ribbon stacking configuration A-5

Figure A-3. Container block stacking A-5

Figure A-4. Container turret stacking (two-high) A-6

Figure A-5. Container turret stacking (three-high) A-6

Figure A-6. Container-on-chassis marshaling system A-7

Figure A-7. Cluster plan for front loader turret stacking of 20-foot containers (50-foot working
aisles) A-8

Figure A-8. Cluster plan for front loader turret stacking of 40-foot containers (70-foot working
aisles) A-9

Figure A-9. Cluster plan for side loader turret stacking of 20-foot containers (15-foot working
aisles) A-9

Figure A-10. Cluster plan for side loader turret stacking of 40-foot containers (15-foot working
aisles) A-10

Figure A-11. 20-foot container cluster A-10

Figure A-12. Traffic pattern in on-chassis marshaling area A-11

Figure A-13. Notional marshaling area A-11

Figure A-14. Suggest traffic flow in permanent terminal marshaling area A-13

Figure A-15. Procedures for marshaling, loading, and unloading containers for rail movements
when rail facilities are not part of or adjacent to the marshaling yard A-14

Figure A-16. Suggested design for a security storage area A-16

Figure A-17. Sample layout plan for container space requirements in a marshaling yard A-19

Figure A-18. Partial layout plan for container space requirements in a marshaling yard A-20

Tables

Table 3-1. Determining factors for number of barges, tugboats, or craft 3-5

Table 3-2. Daily IWWS capacity 3-5

Table B-1. Formula to determine diameter of anchorage site for a ship B-1

Table B-2. Terminal discharge capacity B-2

This page intentionally left blank.

Preface

This Army techniques publication (ATP) introduces Army expeditionary intermodal operations as a capability and describes the roles, responsibilities, and command relationships for planning, executing, and supporting Army terminal operations in theater. This publication applies to the range of military operations and supports Army doctrine publication (ADP) 3-0 and ADP 4-0.

The principal audience for ATP 4-13 is all members of the profession of arms. Commanders and staffs of Army headquarters serving as joint task force or multinational headquarters should also refer to applicable joint or multinational doctrine concerning the range of military operations and joint or multinational forces. Trainers and educators throughout the Army will also use this publication.

Commanders, staffs, and subordinates must ensure that their decisions and actions comply with applicable United States, international, and, in some cases host-nation laws and regulations. Commanders at all levels will ensure that their Soldiers operate in accordance with the law of armed conflict and the rules of engagement. (See FM 6-27/MCTP 11-10C.)

ATP 4-13 uses joint terms where applicable. Selected joint and Army terms and definitions appear in both the glossary and the text. Terms for which ATP 4-13 is the proponent publication (the authority) are italicized in the text and are marked with an asterisk (*) in the glossary. When first defined in the text, terms and definitions for which ATP 4-13 is the proponent publication are boldfaced and italicized, and definitions are boldfaced. For other definitions shown in the text, the term is italicized, and the number of the proponent publication follows the definition.

ATP 4-13 applies to the Active Army, Army National Guard/Army National Guard of the United States, and the United States Army Reserve unless otherwise stated.

The proponent of ATP 4-13 is the United States Army Combined Arms Support Command. The preparing agency is the Deployment Process Modernization Office, United States Army Combined Arms Support Command. Send comments and recommendations on DA Form 2028 *(Recommended Changes to Publications and Blank Forms)* to Commander, United States Army Combined Arms Support Command, ATTN: ATCL-TDID (ATP 4-13), 2221 A Ave, Building 5020, Fort Lee, VA 23801-1809 or submit an electronic DA Form 2028 by email to: usarmy.lee.tradoc.mbx.lee-cascom-doctrine@army.mil

This page intentionally left blank.

Introduction

ATP 4-13 provides the framework for commanders and staffs at all levels on the employment of Army expeditionary intermodal capability to include aerial and seaport operations. The fundamentals of Army expeditionary intermodal operations and general terminal operation techniques tie together various transportation competencies to enhance deployment, redeployment, and distribution operations for the end-to-end movement of personnel, equipment, or forces. ATP 4-13 uses the theater environment as the focus of organizations, events, and activities that are integral to plan and execute the terminal operations that enable expeditionary intermodal operations.

Other intermodal components and transportation competencies that support Army expeditionary intermodal operations are covered in the following doctrinal publications:

- ADP 4-0.
- ATP 3-35.
- ATP 4-02.1.
- ATP 4-11.
- ATP 4-12.
- ATP 4-14.
- ATP 4-15.
- ATP 4-16.
- ATP 4-48.

Though this ATP addresses the general execution of a terminal operations mission, the specific execution of procedures is dependent on the situation or environment. ADP 3-0 states that no two operational environments are the same, and an operational environment can consist of many relationships and interactions among interrelated variables. An operational environment perpetually shifts due to different actor or audience types interpreting messages in different ways. Consequently, ATP 4-13 provides a foundation for commanders to tailor terminal operations as necessary to meet the demands of any operational environment.

ATP 4-13 contains seven chapters and two appendices:

Chapter 1 discusses intermodal operations and their components and introduces Army expeditionary intermodal operations as a capability.

Chapter 2 provides an overview of Army organizational roles during intermodal operations.

Chapter 3 discusses planning, opening, and operating land terminals. The chapter also details the types of land terminals and the responsibilities involved in operating land terminals.

Chapter 4 discusses planning, opening, and operating air terminals. The chapter also details the responsibilities of units involved in operating air terminals.

Chapter 5 discusses planning, opening, and operating maritime terminals. The chapter also details the responsibilities of units involved in operating maritime terminals.

Chapter 6 discusses planning, opening, and operations during logistics over-the-shore activities.

Chapter 7 discusses the transportation brigade expeditionary and its role in supporting expeditionary intermodal operations.

Appendix A provides details on how to establish and operate a marshaling yard. The appendix also discusses space requirements and how to configure containers in a marshaling yard.

Appendix B provides an overview on terminal throughput capacity by type.

This page intentionally left blank.

Chapter 1

Intermodal Operations Overview

This chapter defines intermodal operations and discusses the components of intermodal operations and how they support multidomain operations.

INTERMODAL OPERATIONS

1-1. *Intermodal operations* are the process of using multiple modes (air, sea, highway, rail,) and conveyances (truck, barge, containers, pallets) to move troops, supplies and equipment through expeditionary entry points and the network of specialized transportation nodes to sustain land forces (ADP 4-0). They include the movement of cargo and personnel using two or more transportation modes (surface and air) from point of origin to destination to reduce cargo handling and thereby speed delivery. These operations use movement control to balance requirements against capabilities and capacities to synchronize terminal and mode operations, ensuring an uninterrupted flow through the transportation system. Intermodal operations consist of the facilities, transportation assets, and materials handling equipment (MHE) required to support the deployment and distribution enterprise. Terminal operations and container management are included under this function (FM 4-0). Intermodal operations in the Army incorporate multimodal resources using deployed or commercial assets during contingency operations to support multidomain operations. Intermodal operations provide flexibility for the combatant commander (CCDR) to deploy, employ, and sustain land forces to extend operational reach, ensure freedom of action, and prolong endurance during large-scale combat operations. Figure 1-1 depicts an example of intermodal operations.

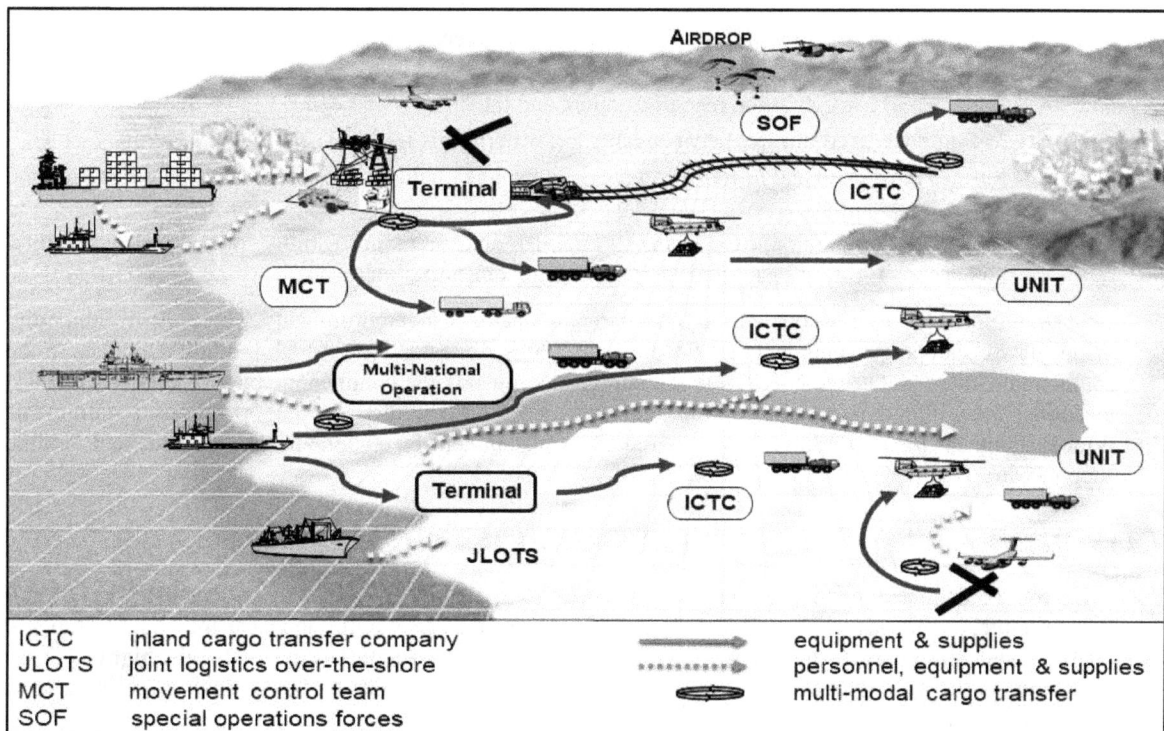

Figure 1-1. Intermodal operational capabilities

1-2. Intermodal operations are multi-faceted operations. They use terminal and mode operations as well as movement control to balance transportation requirements and capabilities against terminal capacities, synchronizing and ensuring an uninterrupted flow of personnel and cargo during large-scale combat operations. At intermodal nodes such as aerial ports of debarkation (APODs) and seaports of debarkation (SPODs), intermodal platforms (for example, containers, 463L pallets, flatracks, cargo, and unit equipment) are offloaded, checked-in, and transferred to a holding area for follow-on movement by transportation assets (truck, rail, watercraft, barge, or air). Service automated information systems provide visibility of transfer actions, including creation of transportation and shipping documentation and radio frequency identification (RFID) tags. Onward movement is arranged, and Service-specific documentation is prepared for the next mode of transportation (see ATP 4-16 for more information on movement control). Intermodal operations enable deployment, movement, and sustainment operations from mobile sea bases, intermediate support bases, seaports, aerial ports, and inland water and land-based terminals, including austere and degraded access nodes and ports of entry in all operational environments.

1-3.   Conducting intermodal operations presents specific challenges in port entry operations. Every host nation (HN) has its own laws and regulations. It is important for commanders and staffs to clearly understand and consider these national agreements and caveats, which may affect the employment and use of HN capabilities. They should consider initiating dialog regarding national agreements and caveats through early planning conferences with partners. These planning conferences may reveal undeclared caveats which may require additional operational resources or sustainment accommodations not previously planned or considered. Operational planners will have realistic views on how to establish intermodal operations in that specific foreign country.

1-4.   Intermodal operations are built on unified action and a significant reliance on multinational partnerships. These multinational partnerships enhance and expand the flexibility and mobility of multinational capabilities to respond to developing crises in a timely manner. A fundamental principle of multinational maritime operations is that Army and multinational partners should be capable of being readily deployed globally to display the flexibility and mobility of multinational maritime power.

1-5.   Through a coordinated process, intermodal operations integrate the tasks and systems that operate water, rail, air, and land terminals, synchronizing those functions with the joint deployment and distribution enterprise and the CCDRs concept of support. Additionally, intermodal operations enable—

- The transfer of forces and cargo from nodes and between movement modes in a manner that meets the commander's operational requirements.
- Rapid cargo configuration, reconfiguration, and transfer.
- Multimodal cargo transfer between land, air, waterborne, rail, and sea-based platforms.
- Surface movement and in-transit support of personnel including combat-ready force elements and civilian populations.
- Transportation automation systems to establish and maintain visibility, provide accountability, and enable in-transit visibility (ITV) to manage transportation assets.

1-6.   Intermodal operations can be employed in all operational environments where Army and joint maneuver forces operate. This includes the ability to move from, to, and through multimodal nodes in the continental United States (CONUS), forward stationed at outside the continental United States (OCONUS) locations, and within theaters of operation. During support to large-scale combat operations, requirements from sustainment forces are increased, thus requiring sustainment formations to provide agile and flexible support when conducting multimodal operations.

COMPONENTS OF INTERMODAL OPERATIONS

1-7.   Intermodal operations take into consideration the joint deployment and distribution enterprise, theater infrastructure, command and control of sustainment units, and the use of multimodal capabilities. *Multimodal* **is the movement of cargo and personnel using two or more transportation methods (air, highway, rail, sea) from point of origin to destination.** Both air and surface modes of transportation are integral to intermodal operations.

INTERMODAL LIFT CAPABILITIES

1-8. Department of Defense (DOD) strategic airlift and sealift systems are intermodal capabilities that facilitate the rapid movement of personnel, equipment, and supplies with minimum disruption to the deployment and distribution flow. They enable rapid offload for transshipment of cargo and equipment for transport by air, water, and ground transportation capabilities for onward movement. The Joint Operation Planning and Execution System is the system of record for coordinating and enabling these movements.

1-9. The DOD airlift system uses both military aircraft and contracted commercial aircraft that can be configured to rapidly load equipment using intermodal platforms such as roll-on/roll-off (RO/RO) ramps and standard 463L pallet systems.

1-10. The DOD sealift system is designed to provide rapid support using government-owned and chartered vessels. United States Transportation Command (USTRANSCOM) uses a significant number of RO/RO vessels with large cargo capacities, vehicle loading ramps, and rapid loading and discharge capabilities as well as commercial containerships to provide intermodal services to convey unit equipment, wheeled and tracked vehicles, containers, flatracks, and supplies.

1-11. The surface transportation system uses highway, rail, and inland waterway systems (IWWSs) to move materiel to aerial ports of embarkation (APOEs) or seaports of embarkation (SPOEs) for loading onto intertheater strategic airlift and sealift assets. USTRANSCOM coordinates with installations to provide intermodal services that include—

- Use of containers and flatracks.
- Line haul transportation to preposition containers at installations and distribution centers.
- Moving containers and unit equipment to SPOEs.
- Port and terminal services.
- Booking space aboard carrier's vessels.
- Offload operations at APODs and SPODs in support of reception, staging, and onward movement.

1-12. Rolling stock can be loaded directly onto railcars to facilitate fast loading at installations and discharge at ports of embarkation during deployment operations. Intermodal containers can be quickly loaded or unloaded from railcars using container handling equipment (CHE) or gantry cranes.

1-13. Containers moved by highway can proceed directly to pier-side for loading aboard containerships using container terminal gantry cranes or ship's cranes. They may also be offloaded from the chassis by specially designed CHE and positioned in the terminal's container yard for loading aboard ships. Containers moved by railcar, which normally do not have direct pier-side access, require CHE for off-load and transfer to pier-side.

1-14. In the surface transportation system, strategic sealift is linked to inland transportation (highway, rail or waterways) through port and water terminal systems (SPOEs/SPODs) to provide for a smooth, seamless flow of equipment and materiel from mode to mode. Strategic sealift vessels provide the primary means of lift for initial unit deployment of unit equipment (for example, tanks, towed artillery, armored infantry fighting vehicles, and rolling stock). Containerships are the ideal means of transport for sustainment, ammunition, and resupply.

1-15. In the air transportation system, strategic airlift is linked to land transportation (highway or rail) through port and air terminal systems (APOE/APOD) to provide a smooth transfer of equipment and materiel from mode to mode. Military assets are configured to allow equipment to be driven on or off the aircraft and rapidly loaded or unloaded using the 463L pallet system. The 463L pallet and netting are in high demand during deployment and large-scale combat operations. While it is preferred that 463L assets remain in the airlift system, some 463L pallets may remain configured (and later returned) for onward movement to meet the commanders' objectives and priorities.

MODE OPERATIONS

1-16. *Mode operations* are the execution of movements using various conveyances (truck, lighterage, railcar, and aircraft) to transport cargo (ADP 4-0). There are two transportation modes of operation available to support military operations—surface and air. The surface mode includes motor, water, and rail. The air mode consists of fixed-wing and rotary-wing aircraft. The mode selection is influenced by the type of military operation and transportation asset availability, capability, and capacity. Mode operations also include the administrative, maintenance, and security tasks associated with the operation of the conveyances. For more information see ATP 3-35, ATP 4-11, ATP 4-14, ATP 4-15, and ATP 4-16.

TERMINAL OPERATIONS

1-17. Terminal operations include the reception, processing, and staging of passengers; the receipt, transit, storage, and marshaling of cargo; the loading and unloading of modes of transport conveyances; and the manifesting and forwarding of cargo and passengers to destination. Terminal operations involve transferring personnel or transshipping cargo between two or more modes of transportation to complete movement to final destination.

1-18. Terminals are essential nodes in the total distribution network for deployment, redeployment, and sustainment that support the commander's concept of operations. When linked by modes of transport (air, highway, rail, and water), terminals define the transportation structure. Large-scale combat operations require the early identification and establishment of terminals to support early entry operations.

1-19. The assignment of organizations with personnel, unit equipment, CHE, and MHE sufficient to meet the workload requirements at each terminal is crucial to the execution of terminal operations. Staff planners at all levels must provide for the adequate manning of terminals. They must plan for workable solutions in cases where terminal facilities are insufficient for necessary throughput. In addition, automated information systems capable of supporting ITV requirements for the movement of the personnel, equipment, and materiel moving through terminals are important. These systems provide the combatant commands (CCMDs) with information pertaining to location and final destination of all cargo.

1-20. The loss of space-based communications is a concern for Army forces conducting terminal or port operations. Whether the interruption of the communications is caused by enemy action against satellites or through the use of intermittent jamming, the resulting black-out will require forces to adapt and adjust until the capability is restored. Short-term losses or disruptions of satellite communications can be mitigated through alternative communications methods and courier networks. The United States (U.S.) Air Force has an entire squadron dedicated to observing and forecasting space weather activity and anticipating disruptions to communications. The Staff Weather Officer can assist with forecasting anticipated effects to space-based communications.

1-21. During support to large-scale combat, movement control supports conflict resolution by planning, routing, and scheduling movements through multiple access points including degraded and austere environments. Movement control also provides ITV during the development of combat power, aiding in the coordination of combat capability to defeat the enemy.

CONTAINER MANAGEMENT

1-22. *Container management* is the process of establishing and maintaining visibility and accountability of all cargo containers moving within the Defense Transportation System (ADP 4-0). Effective container management involves, at minimum, coordination between Military Surface Deployment and Distribution Command (SDDC), Air Mobility Command (AMC), Defense Logistics Agency, and the theater sustainment command (TSC) distribution management center and movement control battalion (MCB). This coordination enables the effective management, deployment, and distribution of containers of all types within the intertheater and intratheater segments of the distribution pipeline. Containers facilitate and optimize cargo-carrying capabilities and capacities of intermodal transport. When required for tactical unit distribution, joint modular intermodal containers may provide a smaller containerized ability to move small munitions, parts, and supplies. ATP 4-12 provides additional information on container management.

TERMINAL TYPES

1-23. Terminals are generally classified based on the following attributes: physical characteristics, type of transport assets used, type of cargo handled, and the methods used for cargo handling. The primary types of terminals operated by the Army include land, air, and maritime. Determining the cargo capacity of terminals is critical during the planning process for military operations.

LAND TERMINALS

1-24. Land terminals include inland water, rail, highway, or petroleum terminals. Petroleum terminal operations are covered in ATP 4-43. Centralized receiving and shipping points (CRSPs) and trailer transfer points (TTPs) are also considered land terminals. Land terminals are established at points along air, rail, river, canal, pipeline, and motor transport lines of communications (LOCs) to provide for the transshipment of cargo and personnel carried by these modes.

1-25. Highway or motor transport terminals are normally located at both ends of a line-haul operation. They form the connecting link between local hauls and the line-haul service. They may also be located at intermediate points along the line-haul route where terrain necessitates a change in type of carrier. Inland cargo transfer companies (ICTCs) provide cargo-handling service at motor transport terminals. See ATP 4-11 for more information on motor transport operations and ATP 4-16 for more information on movement control.

1-26. Inland water terminals are key enablers or links between modes when terrain and operational requirements cause a change in type of conveyance. Inland waterways include all rivers, lakes, inland channels, canals deep enough for waterborne traffic, and protected tidal waters. They include the locks, dams, bridges, and other structures that contribute to or effect movement of vessels carrying passengers and freight. ICTCs provide cargo-handling service at inland waterway terminals (IWWTs).

1-27. Rail terminals enable intermodal operations and can be critical nodes in establishing distribution throughput. Each rail terminal type can differ in size, location, and primary function, but usually retains the common functions of loading and unloading of cargo or personnel. Rail terminals include rail yards, rail heads, freight stations, passenger stations, and repair and service facilities. Most rail terminals are located at the start and end of rail lines. Well-established rail systems have one or more rail yards between the start and end of a line. Railheads are located on the forward end of a military railway where personnel, supplies, and equipment are transferred to other modes of transportation for further movement forward. See ATP 4-14 for more information on rail operations.

AIR TERMINALS

1-28. An *air terminal* is a facility on an airfield that functions as an air transportation hub and accommodates the loading and unloading of airlift aircraft and the in-transit processing of traffic (JP 3-36). Air cargo transfer operations occur at Air Force and Army air terminals. Examples of traffic moved through an air terminal include passengers, unit equipment, sustainment cargo, and mail. The ICTC loads and unloads aircraft, documents cargo moving through the terminal, and operates cargo segregation and temporary holding facilities. Movement control teams (MCTs) located at the terminal coordinate the flow of cargo and passengers.

1-29. Air terminals may be established on airfields of a military Service other than the U.S Air Force. Certain airfields are designated as aerial ports for strategic air movements supporting deployment, redeployment, and sustainment operations. An *aerial port* is an airfield that has been designated for the sustained air movement of personnel and materiel, as well as an authorized port for entrance into or departure from the country where located (JP 3-36). Aerial ports provide the most expeditious method for rapid force deployment and normally serve as a link to land transportation systems in the theater. See chapter 3 for further details on air terminal operations.

MARITIME TERMINALS

1-30. Maritime terminals can be located at permanent port, unimproved port, or bare-beach facilities. Ports can be classified as general cargo (does not include ammunition and bulk liquids), container, RO/RO, or a combination of capabilities. Bare-beach facilities or logistics over-the-shore (LOTS) operations provide expeditionary access nearer to tactical assembly areas. LOTS operations provide a means for loading and unloading ships without the benefit of deep draft-capable, fixed port facilities, or as a means of moving forces closer to tactical assembly areas dependent on threat force capabilities. See chapter 6 for additional information on LOTS operations.

TERMINAL PLANNING CONSIDERATIONS

1-31. When planning support for military operations, planners and operators at the joint and Service level must consider the diversity and challenges of managing and operating Army-controlled terminals and commercial or HN-controlled terminals where an Army presence is not available. Planners leverage bilateral or multilateral diplomatic agreements to access ports, terminals, airfields, and bases within the area of responsibility to support future military contingency operations. Land, air, and maritime terminals will be required in most force projection operations; their capabilities to support the various modes of transportation are vital and foremost to planning considerations. Terminal planning is part of the transportation plan developed to support the CCDR's operations plan.

1-32. Early planning considerations include evaluating all aspects of operating Army terminals. A physical assessment is not always necessary if adequate information is available during pre-deployment planning about the available facilities, provisions for HN support, and the immediate area of operations (AO) to minimize the risk to operations. Acquisition and cross-servicing agreements are policy and procedural tools that are often used to facilitate Army expeditionary intermodal operations and may be used to expedite terminal support planning. When executed, these agreements provide for greater flexibility to meet requirements for terminal operations and may expedite or eliminate some of the issues associated with articulating requirements and having to establish contracts or outsource for support. Acquisition and cross-servicing agreement authorities and acquisition-only authority agreements for logistic support, supplies, and services from eligible countries and international organizations are verified in the Acquisition and Cross-Servicing Agreement Authorities Global Automated Tracking and Reporting System. When a physical assessment is required, it should gather adequate information to—

- Acquaint the commander with the layout.
- Recon the area to determine the physical layout and placement of units, support equipment, and automatic identification technology hardware.
- Determine if the distribution infrastructure is capable of supporting the stated mission.
- Determine the availability of resources (or recommended additional assets) required to accomplish the defined tasks.
- Plan for HN support and contract labor.

1-33. Terminal planning includes—

- Estimating the existing terminal throughput capacity. This is the estimated total tonnage and numbers of personnel and containers that can be received, processed, and cleared through the terminal in a day.
- Computing the terminal workload needed to support the operation. The workload is expressed as numbers of personnel, vehicles, twenty-foot equivalent unit (TEU) containers, and short tons (STONs) for non-containerized cargo. This computation includes the total tonnage and numbers of personnel and containers that must be received, processed, and cleared through the terminal.
- Determining repair and rehabilitation costs of existing facilities or new construction needed to increase existing terminal capacity to equal computed terminal workload. (Existing terminal capacity maybe insufficient to support the operational workload.)
- Estimating the CHE and MHE needed to process the required workload, including equipment such as pallets, forklifts, tugs, barges, and cranes, and the operators required to operate them.
- Estimating the units, personnel, civilian augmentation support, HN support, and supervisory and command requirements needed to operate the terminal.
- Identifying and estimating security personnel requirements in case military police or HN support is not available.

1-34. Once the AO is assigned and the mission is understood, the commander should consider the following planning factors to establish successful terminal operations:

- Physical characteristics and layout of the terminal area:

- Physical restrictions of working space and parking space (may impact capacity).
- Availability of hard surfaces in transfer areas.
- Existing facilities for storage and maintenance of MHE and other equipment.
- Proximity of exit routes to transfer points.
- Distances between loading and unloading points and temporary holding areas.
- Security and safety standoff distances.

- Transportation equipment planning factors:
  - Number of commercial carriers that can be handled simultaneously.
  - Delivery turnaround times.
  - Loading and unloading rates for various types of transportation.
  - Effects of size and maneuverability of carriers at transfer points within the terminal.
  - Types and number of CHE and MHE required.

- Types of cargo to be handled:
  - Size and type of packaging.
  - Average weights of cargo units.
  - Requirements to break down into smaller lots or consolidate for reloading.
  - Fragile and perishable cargo.
  - Hazardous cargo, ammunition net explosive weight and compatibility.

- Automated identification technology requirements to provide ITV:
  - Hardware and software capabilities required at permanent and temporary terminals.
  - Interrogator data collection points.
  - Number of interrogators required to capture and report ITV data.
  - ITV plan is consistent with and supports the theater ITV plan.

- Temporary in-transit storage facilities:
  - Type and size facilities required.
  - Security protection requirements.
  - Distances from transshipment or transloading points.
  - Documentation requirements.
  - Requirements for MHE in storage areas.
  - Capacity to transfer cargo from point of discharge to storage.

- Policy and procedures:
  - Standard operating procedures and guidelines for terminal operations that ensure compliance with safety, documentation, communications, ITV, distribution, and equipment maintenance requirements.
  - Policy and guidelines to comply with all applicable international, state, local, standardization agreement, North Atlantic Treaty Organization, and HN laws and regulations to include environmental regulations. See JP 3-34, 32 CFR Part 651, AR 200-1, and ATP 3-34.5.
  - Enter bilateral acquisition and cross-servicing agreements for the provision of logistics support, supplies, and services.
  - Plan, provide, and manage operational contract support to procure supplies and services.

- Other considerations:
  - Anti-terrorism and force protection considerations.
  - Procedures for complying with all applicable international, local, and HN environmental regulations, including oil spill contingency planning, waste disposal, and site-specific environmental concerns. See DODM 4715.05, Vol. 1 and TM 3-34.56.
  - Environmental considerations/factors may include, but are not limited to, environmental compliance, pollution prevention, conservation, historical and cultural property protection,

and flora and fauna protection. (Including cultural, historical, and natural resources. See DODI 4715.22, AR 200-1, and ATP 3-34.5.

- Environmental health site assessment
- Environmental baseline survey.
- Weather and environment.
- Administration and communications.
- Refueling.
- Dining and billeting.
- Latrines.
- Laundry and showers.
- Vehicle recovery and maintenance.
- Medical.

1-35. An environmental baseline survey is a multi-disciplinary site survey conducted before or in the initial stage of a joint operational deployment. (JP 3-34) If the tactical situation permits, commanders must conduct or direct an environmental baseline survey before occupying any of these sites. The environmental baseline survey will help determine previous site usage, hazards on the site, and the potential for hazards generated from areas surrounding the site. Hazards are those things that are generated as a result of military operations and include both those presented to personnel occupying the site and to the surrounding indigenous populations and institutions.

PORT OPENING

1-36. *Port opening* is the ability to establish, initially operate and facilitate throughput for ports of debarkation to support unified land operations (ADP 4-0). Port opening is a subordinate function of theater opening and port opening elements should precede the arrival of deploying combat forces (except in the case of forcible entry). The port opening process is complete when the port of debarkation (POD) and supporting infrastructure is established to meet the desired operating capacity for that node. Supporting infrastructure can include the transportation required to support port clearance of cargo and personnel, holding areas for all classes of supply, and the proper ITV systems established to facilitate force tracking and end-to-end distribution.

1-37. Strategic port opening is a joint process that is normally performed by the CCDR and supported by USTRANSCOM. Austere or degraded port opening may require a mix of joint and Service-specific forces to accomplish the mission. For example, depending on mission variables, a CCDR could use one of the following packages to open an APOD (also see Figure 1-2 for a notional joint early entry structure for port opening):

- AMC's contingency response group and an Army arrival/departure airfield control group (A/DACG).
- USTRANSCOM's joint task force port opening (JTF-PO) APOD element.

1-38. To open a SPOD, the CCDR could use one of the below force packages (depending on mission variables):

- SDDC elements contracting stevedore and other terminal functions.
- USTRANSCOM's JTF-PO SPOD element.
- SDDC elements in conjunction with transportation brigade expeditionary (TBX) or attached subordinate units.

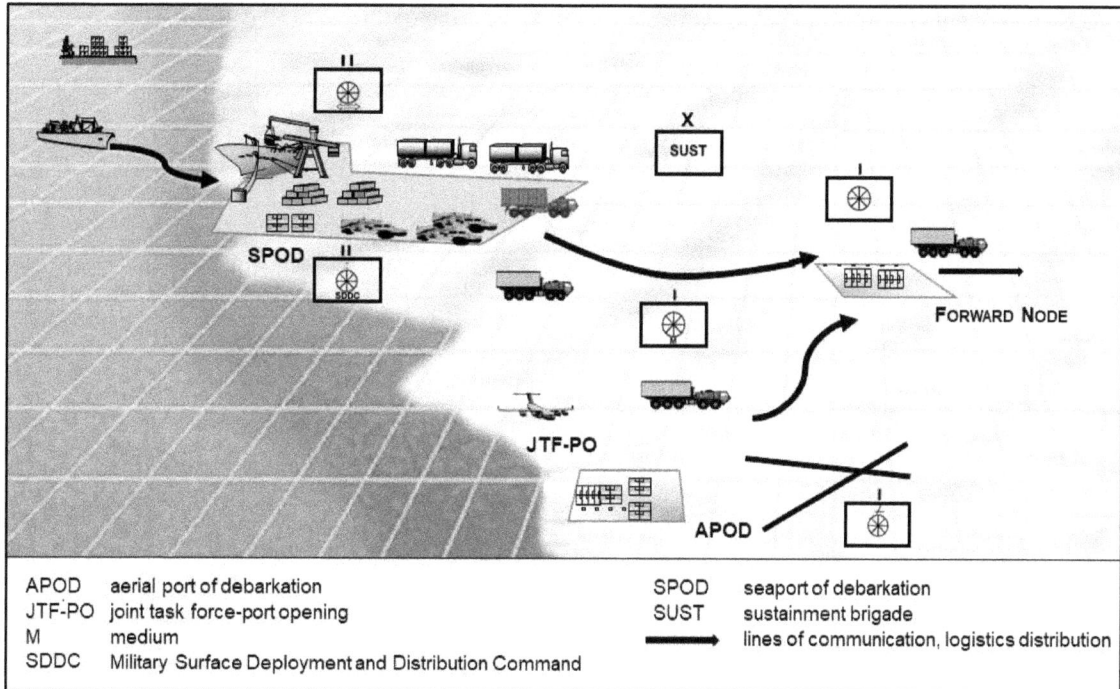

APOD   aerial port of debarkation   SPOD   seaport of debarkation
JTF-PO joint task force-port opening   SUST   sustainment brigade
M   medium   ———►   lines of communication, logistics distribution
SDDC   Military Surface Deployment and Distribution Command

Figure 1-2. Notional joint early entry structure for port opening

1-39. The TSC is responsible for opening the theater (which port opening supports). Theater opening also includes—

- Establishing communications.
- Intelligence.
- Civil-military operations.
- Services.
- Human resources.
- Financial management.
- Coordinating with the theater medical command for Army Health System support.
- Engineering.
- Movement (air, land, water transport, inland terminal operations).
- Materiel management.
- Maintenance.
- Coordinating operational contract support coordination.

1-40. The TSC is the vital link for successful theater opening and must establish command and control infrastructure early in the theater opening process. This supports not only port opening, but reception, staging, onward movement, and integration (RSOI) and initial distribution operations (at forward distribution nodes) as well. See FM 4-0 for more information. Establishing this infrastructure early is especially critical when supporting an operation that leaves little time for the theater to build up (for example, a rapid deployment to support a foreign humanitarian assistance and disaster relief effort). Regardless of whether USTRANSCOM or TSC forces open a port, they will require elements of a sustainment brigade, combat sustainment support battalion (CSSB), expeditionary terminal operations element, or a TBX to provide command and control for the distribution and sustainment operations supporting port and theater opening. If HN or contract truck transportation is not available, then light or medium truck transportation will be required. A sustainment brigade (-) or a CSSB can provide the core command and control structure for sustainment early entry operations to support port clearance, initial distribution operations, initial movement control with highway regulation, and RSOI while setting the conditions to expand the sustainment footprint. Multiple ports can open within a short window of one another (such as an APOD to bring in troops and an SPOD for equipment), so it is important that TSC units arrive early in the force flow to coordinate and synchronize the various port activities and theater distribution plans.

Chapter 2

Organizations, Roles, and Functions

Intermodal operations, multimodal capabilities, and the required infrastructure to support multidomain operations were discussed in chapter 1. This chapter provides an overview of Army organizations, roles, and functions integral to intermodal operations. This chapter includes a discussion of joint force and unified action partner organizations.

ARMY ORGANIZATIONS

2-1. To counter threats and protect national interests worldwide, the Armed Forces of the U.S operate as a joint force in unified action. Army forces, conduct operations in support of the joint force with multinational allies and unified action partners in coordination with other agencies and organizations. The Army provides unique capabilities such as terminal operations, movement control and intermodal operations to multidomain operations in support of unified action.

ARMY SERVICE COMPONENT COMMAND

2-2. The Army Service component command (ASCC) is the command responsible for recommendations to the joint force commander on the allocation and employment of Army forces within a CCMD (JP 3-31). ASCC is a Service role. The ASCC is responsible for RSOI and for managing redeployment of forces. Additionally, Service component commanders retain responsibility for certain Service-specific functions and other matters affecting their forces, including internal administration, personnel support training, sustainment (with some exceptions), and Service intelligence operations. There can be only one ASCC within the CCMD. This is the primary role of the theater Army.

2-3. The theater Army is the ASCC assigned to the CCMD. A major task of the ASCC is to set the theater. The phrase "set the theater" or "setting the theater" is used to capture the broad range of functions and tasks conducted to shape the operational area and establish the conditions across an AOR that enable the execution of the theater strategic plans as established by the CCMD campaign plan. The purpose of setting the theater is to shape conditions to gain access required to facilitate future military operations, sustain Army and joint forces within an AOR, and facilitate the successful execution of the CCMD campaign plan and other theater strategic plans. See FM 3-94 and ATP 3-93 for more information on ASCCs. Part of setting the theater is to open LOCs to build an intermodal distribution network that includes—

- Terminals and facilities required to move, maintain, and sustain theater forces.
- Control of RSOI for Army forces in the AO.
- APOEs and APODs.
- SPOEs and SPODs.
- Water, rail, and route networks.
- HN resources.

2-4. Key responsibilities of the ASCC include RSOI and sustainment. The ASCC may also have many lead Service responsibilities, which entail common-user logistics support to other Services, multinational forces, or unified action partners. The primary terminal operations-related functions of the ASCC are as follows:

- Equip, train, and employ U.S. Army units for LOTS operations in coordination with U.S. Marine Corps and Navy units.
- Provide management of overland petroleum support, including IWWS, to U.S. land-based forces of all DOD components.
- Operate Army watercraft along intratheater sea LOCs and IWWS.

- Operate some or all maritime terminals in the theater in coordination with the single port manager, SDDC.
- Provide pipeline fuel support.
- Provide engineer support for inland distribution network (highways and bridges).
- Provide rotary-wing common-user support.
- Provide logistic support to multinational commands or HNs for specific support, as directed.

UNITED STATES ARMY MATERIEL COMMAND

2-5.   United States Army Materiel Command (USAMC) is designated an Army Command by the Secretary of the Army to manage the Army's logistics mobilization and contingency capability and capacity. USAMC equips and sustains the Army. USAMC provides technology and acquisition support in support of multidomain operations to ensure dominant land force capability for U.S. forces and unified action partners.

MILITARY SURFACE DEPLOYMENT AND DISTRIBUTION COMMAND

2-6.   SDDC is an operational-level Army force designated by the Secretary of the Army as the ASCC of USTRANSCOM and a major subordinate command of USAMC. SDDC is responsible for providing global deployment and distribution planning, operations, and systems capabilities and for facilitating global traffic management support to all joint, multinational, and interagency elements. SDDC—

- Provides DOD deployment and distribution management services for freight, unit, and personal property movements worldwide.
- Provides worldwide coordination from origin to destination for surface traffic management support. This includes coordinating surface and multimodal transportation contracted functions for all DOD (and other U.S. Government entities, as authorized, and designated multinational and interagency elements) freight and unit movements and providing worldwide management services for DOD personal property.
- Plans and executes oversight of command acquisitions for transportation services to support CCMD requirements for enduring and contingency operations and infrastructure.
- Coordinates with appropriate acquisition authorities and is the sole DOD negotiator worldwide with commercial service providers on rates and other matters incidental to transportation and storage services of the personal property of all DOD personnel.
- Manages and arranges for the operation of common-user ocean terminals in CONUS and operates or arranges for the operation of OCONUS ocean terminals under agreements with appropriate commanders and civil authorities.
- Coordinates with Military Sealift Command (MSC) to book cargo aboard liner service vessels, to include multimodal shipments in accordance with (IAW) contractual agreements and provides appropriate support to movements occurring on government ships.
- Coordinates with CCDRs to perform water terminal clearance authority functions.
- Develops, operates, and maintains an integrated transportation information system to support the transportation mission and provides traffic management information and data for DOD components.
- As DOD's global container manager, provides operational management of defense intermodal common-user containers and oversees and operates a worldwide DOD surface container management system.
- Controls, manages, and maintains the Defense Freight Railway Interchange Fleet.
- Performs business intelligence functions to facilitate studies and analysis of transportation requirements, capabilities, organizations, operations, planning, effectiveness, and economies, and recommends improvements for DOD implementation.
- Participates in the planning cycle for overseas deployment, training exercises, and command post exercises directed by the Joint Chiefs of Staff and recommends corrective actions when military or commercial transportation assets or procedures cannot support mission accomplishment.

- Establishes standards for, facilitates, and validates training of Army Active and Reserve Component strategic mobility forces to ensure capable and ready forces to meet SDDC missions and operation plan support.
- Coordinates with DOD components to maintain joint Service publications governing installation shipping and receiving capabilities.

2-7. The transportation surface brigade is an Active Component SDDC Table of Distribution and Allowance (TDA) headquarters responsible for command and staff oversight of assigned water terminals. Transportation surface groups perform staff functions and management in support of subordinate transportation units.

2-8. The transportation battalion is an Active Component SDDC TDA unit under the command of the transportation surface group. It is designed to conduct surface deployment, distribution, and water terminal port operations directly supporting units in its assigned AO. The transportation battalion—

- Plans, establishes, and conducts port operations to include cargo reception, staging, load planning, and vessel load/discharge operations.
- Commands terminal management teams engaged in managing contract operations at a SPOE or SPOD.
- Transitions from command and staff oversight of Army table of organization and equipment terminal operating units to managing contract capabilities at SPODs or APODs.
- Provides a port common operational picture.
- Serves as a single port manager of a strategic seaport.
- Supports port opening operations.

2-9. The Deployment Support Command is a United States Army Reserve (USAR) TDA headquarters with the mission to command and provide staff oversight of SDDC-assigned or attached Army Reserve units. It provides standardized training and readiness oversight to all Army units engaged in water terminal, deployment and distribution support, container management, and movement control operations. It is under the operational control of SDDC and administrative control of the TSC.

2-10. The transportation surface brigade is an Active Component or USAR TDA headquarters that commands, controls, and technically supervises assigned or attached SDDC transportation battalions, deployment and distribution support battalions (DDSBs), or transportation terminal battalions engaged in terminal operations, terminal supervision and management operations, and other mobility support operations.

2-11. The DDSB is a USAR TDA headquarters designed to command, control, and technically supervise terminal companies and detachments operating at seaports. Each battalion has three deployment and distribution support teams (DDSTs) and two terminal management teams integral to it. The DDSB—

- Commands and controls DDSTs which provide technical deployment-related support to deploying units worldwide and container management in theater.
- Commands and controls terminal management teams engaged in supervising operations in a SPOE or SPOD.
- Commands and controls other transportation units (terminal operations elements, automated cargo documentation detachments, or seaport operations companies (SOCs) performing terminal operations in a SPOE or SPOD, as necessary.

2-12. Attaching an expeditionary terminal operations element increases the terminal management capability of a DDSB. The addition of the automated cargo documentation team increases the DDSB's berth capability. The exact number of teams in any given DDSB will depend on routine and daily operations in CONUS or OCONUS, as well as theater wartime requirements. When deploying to new port areas, they may be supplemented with teams from other active battalions and backfilled by USAR battalions.

2-13. The DDST assists units with deployment planning and staging and preparing unit equipment and personnel for worldwide movement by surface or air. When deployed to a theater of operations, the DDST will manage, control, and maintain ITV of containers moving in theater. DDSTs can be attached to USAR DDSBs or active transportation battalions. The DDST provides the DDSB or transportation battalion with integral, modular capability to meet deployment support mission requirements and can—

- Provide deployment assistance to the brigade mobility officer or installation transportation officer and the air and seaport operating units.
- Assist units with movement to a designated port of embarkation (POE) or POD.
- Provide deployment support from fort to port through movement planning, preparation, and communication.
- Ensure accuracy of documentation associated with deploying equipment.
- Provide liaison between the port and installation to minimize the frustrated cargo and equipment at the port.
- Ensure the conduct of safe operations (rail load or line haul) through effective management and control.
- Provide technical guidance and assistance to units in preparing, maintaining, and executing movement plans, unit movement data.
- Provide guidance on how to obtain blocking, bracing, packing, chains, and tie-down equipment.
- Inspect equipment to ensure that vehicles are correctly identified, cargo is properly loaded on the vehicles, and no equipment is missing that would impair the loading operations at the port.
- Provide hazardous material-qualified personnel to assist unit hazardous material certifiers in preparation of hazardous cargo documentation.
- Coordinate with U.S. Coast Guard Container Inspection and Training Assistance Team to conduct training and inspect containers and hazardous material cargo at installations for OCONUS deployments; coordinate with the Redeployment Assistance Inspection Detachment Team for U.S. Coast Guard mission support within theater.
- Provide daily situation report, ensure DDSTs or active transportation battalions verify and validate RFID ITV data sent to the integrated data environment/global transportation network convergence system and ITV updates are viewable to the joint deployment and distribution enterprise.
- Provide personnel to monitor and report on container movements in theater.

THEATER SUSTAINMENT COMMAND

2-14. The TSC is the Army's command for the integration and synchronization of sustainment in the area of responsibility. The TSC is assigned to a theater Army and focuses on Title 10 support of Army forces for theater security cooperation and the CCDR's daily operational requirements. The TSC has three operational responsibilities to forces in theater: theater opening, theater distribution, and sustainment (FM 4-0). The core competency of the TSC is command and control of sustainment units executing theater distribution to include terminal operations, movement control, and multimodal operations (JP 4-09).

2-15. The TSC and its subordinate elements coordinate and synchronize movement control while allocating transportation assets to support movement requirements. The TSC or expeditionary sustainment command (ESC) monitors movements at land terminals and inland waterway (IWW) terminals through MCBs. MCBs and MCTs are essential in regulating and controlling movements and ensuring planned schedules are met. The number of MCBs and MCTs can vary depending on size of the operational area. MCBs and MCTs also serve as an interface between operational planners and users.

2-16. The theater petroleum center serves as the senior Army petroleum advisor to the CCMD. The theater petroleum center provides theater strategic through operational planning support to CCMDs, the theater Army, corps, and TSC (ATP 4-43). The theater petroleum center—

- Validates time-phased force and deployment data for petroleum and water support units and command and control elements at the theater Army level and below.
- Determines transportation (intratheater and intertheater) requirements and methodology for multimodal distribution network movement of bulk petroleum and alternative fuels from the point of receipt of product from DLA Energy forward.
- Conducts multimodal distribution network and storage capabilities assessments of HN, allied, and partner nations in concert with theater strategic and operational partners.

EXPEDITIONARY SUSTAINMENT COMMAND

2-17. ESCs are designed to extend the operational reach of the TSC to provide more responsive support to Army forces. The ESC is employed as the forward headquarters and commands attached sustainment units to provide command and control for theater opening, theater distribution, and theater sustainment within the operational support area of the theater. The ESC is responsible for port and terminal operations and RSOI. The ESC normally establishes its command post near the ports of debarkation to control both reception and sustaining operations. This location may be a secure base within the joint operations area or joint security area. See FM 4-0 for additional information on the ESC.

SUSTAINMENT BRIGADE

2-18. The sustainment brigade is a multifunctional logistics headquarters that provides command and control for sustainment units. Sustainment brigades are attached to a sustainment command. The sustainment brigade conducts support operations that include planning, coordinating, and synchronizing sustainment in support of units in theater and corps areas of operations. The sustainment brigades support theater opening, theater distribution, and sustainment. These missions can involve port and inland terminal operations, RSOI, CRSPs and TTPs. Sustainment brigades may be augmented with multifunctional subordinate units to support port clearance and ongoing distribution operations, ensuring timely movement of cargo to and from terminals.

COMPOSITE WATERCRAFT COMPANY

2-19. The composite watercraft company provides command and control, operations planning, and maintenance support for up to 16 Army watercraft detachments performing intra-theater lift, water terminal or harbor operations, waterborne tactical and joint amphibious, riverine, and LOTS operations. These may include a combination of logistics support vessel; landing craft, utility; landing craft, mechanized or maneuver support vessel-light; and small tug detachments.

TRANSPORTATION BRIGADE EXPEDITIONARY

2-20. The TBX provides command and control of Army watercraft and water terminal capabilities and organizations. The TBX deploys to a theater of operations to provide command and control for port opening and operations at inland waterway, bare beach, degraded, and improved sea terminals. The headquarters is organized to provide the ability to rapidly deploy minimum capabilities to meet rapid port opening operations and small-scale contingencies. An expanded discussion on the TBX is available in chapter 7.

MOVEMENT CONTROL BATTALION

2-21. MCBs are usually assigned to TSCs or ESCs conducting movement control operations at the operational or tactical level of warfare. The MCB controls the movement of all U.S. forces, their equipment, materiel, and sustainment into, within, and out of its assigned AO. It commands between four and ten MCTs and is responsible for the execution of the TSC or corps movement program and performance of the transportation system. The MCB provides transportation asset visibility and coordinates the use of common-user transportation assets, intermodal container assets such as International Organization for Standardization containers, 463L pallets, and flat racks. The MCB maintains ITV radio frequency networks to provide ITV updates of unit moves and convoy movements via integrated data environment/global transportation network convergence feeds. Since MCTs are normally positioned at airfields as part of the A/DACG mission, MCBs can be the battalion-level Army command for airfield operations in the theater. MCBs provide a link between the theater strategic, operational, and joint movement community. The MCB also provides interconnectivity with various intermodal LOCs, enhancing theater distribution and deployment operations. See ATP 3-35 for more information on the MCB.

MOVEMENT CONTROL TEAM

2-22. MCTs are attached to the MCB to provide decentralized execution of MCB movement responsibilities throughout a specified AO. MCTs may be employed on an area basis or at critical node to facilitate effective movement control. MCTs have five primary operational functions: provide area support, port (air and sea) support, movement regulating, divisional support, and cargo documentation. The port support mission at an air terminal is part of the A/DACG mission to process inbound and outbound personnel and cargo as well as terminal clearance. The size and role of movement control at a terminal is determined by the tasks to be accomplished. MCTs could be responsible for operating the passenger terminal, air load planning, the inbound and outbound marshaling yards, or arranging transportation to move passengers and equipment. Additionally, elements of the MCT may collocate at the CRSP to provide and maintain ITV and radio frequency networks, convoy assistance, or other transportation services as needed. For more information on MCTs, see ATP 3-35 and ATP 4-16.

COMBAT SUSTAINMENT SUPPORT BATTALION

2-23. The CSSB is a tailored, multifunctional logistics (less medical) organization. It provides flexible and responsive logistics support on an area basis throughout the depth of its assigned AO. The CSSB subordinate elements may consist of functional companies providing supplies, ammunition, fuel, water, transportation, cargo transfer, mortuary affairs, maintenance, and field services; they could be customers of the CRSP as well. The CSSB ensures the CRSP has all required resources and provides the truck transportation necessary for the onward movement of cargo. If transportation requirements exceed the CSSB's capacity or capability, the battalion will request assistance from their sustainment brigade.

INLAND CARGO TRANSFER COMPANY

2-24. The ICTC mission is to discharge, load, and transship cargo at air, land or truck terminals, theater distribution center, and CRSP. ICTCs supplement cargo and supply handling operations to alleviate cargo backlogs and operate cargo marshaling areas as required. An entire ICTC is usually not needed at an air terminal, but rather a squad or platoon-sized element depending on the capacity of the airfield. The ICTC can transship 1,500 STONs of breakbulk cargo or 600 containers at a terminal. Elements of the ICTC perform joint inspections, provide minor maintenance for equipment to be loaded or cleared from the terminal, provide handling of palletized or containerized cargo, and provide transportation to assist with terminal clearance. Normally attached to a CSSB, ICTCs operate on a 24-hour, two-shift basis to operate intermodal terminals in a theater hub or theater distribution center.

ARRIVAL/DEPARTURE AIRFIELD CONTROL GROUP

2-25. The A/DACG is an ad-hoc organization established to control and support the arrival and departure of personnel, equipment, and sustainment cargo at airfields. An MCT and an ICTC typically operate the A/DACG. The base organization of an A/DACG consists of a 21-person MCT and a squad from an ICTC with similar capabilities as the JTF-PO. An MCT acts as the Army liaison with the Air Force to provide a detachment-level command and control structure to process passengers, conduct air load planning, coordinate multimodal cargo transfer and transportation, and conduct cargo documentation to facilitate the onward movement for cargo and passengers. Elements of an ICTC augment an MCT to provide the personnel and equipment to load and off-load aircraft as needed, transport cargo for airfield clearance, process outbound equipment, and provide minor maintenance support. For more information see ATP 3-35.

2-26. Unlike a JTF-PO, units identified to operate the A/DACG will be part of a time-phased force and deployment data and must arrive early enough to establish operations to set the stage for the reception of personnel, equipment, and sustainment cargo. The A/DACG takes direction from the sustainment brigade during the initial phases of port opening until the supporting MCB arrives in theater. Once the supporting MCB establishes operations, it can assume command and control of the A/DACG and coordinate with the sustainment brigade theater movement control element as needed.

TRAILER TRANSFER POINT TEAM

2-27. A TTP team is a table of organization and equipment element that operates a TTP or convoy support center. The TTP team is assigned to a TSC and normally further attached to either a transportation motor transport battalion or a CSSB. There are two sections in the TTP team. The TTP team operations section provides command, control, and supervision of unit movement plans and routine specialized operations. The maintenance section provides field maintenance to organic wheeled vehicles and emergency maintenance for up to 10 percent of transit vehicles.

INLAND WATERWAY ORGANIZATIONS

2-28. Establishing and operating IWWTs requires the same type of organizations and units that establish and conduct operations at maritime terminals. These units must adapt to any theater-specific requirements that determine the scope of each organization's involvement in establishing and operating IWWTs.

TERMINAL BATTALION

2-29. Terminal battalions are assigned to a TSC and are then normally further attached to a sustainment brigade or TBX. The terminal battalion is an echelon above corps asset employed in permanent ports, unimproved ports, and bare beach facilities. It provides command and control for three to seven subordinate transportation companies or equivalent units performing cargo handling operations and boat operations on inland waterways. Subordinate units may include an ICTC, port operations cargo company, harbormaster detachment, and composite watercraft company with assigned or attached Army watercraft.

SEAPORT OPERATIONS COMPANY

2-30. SOCs perform seaport terminal service operations to discharge, and load containerized breakbulk cargo and wheeled or tracked vehicles in permanent seaports or in LOTS operation sites. These units coordinate seaport clearance and onward movement with supporting movement control and motor transport units. A SOC is dependent on combat heavy equipment transporter companies for the relocation of heavy equipment, and medium truck companies for movement of 40ft containers if the marshaling area is located more than 5 miles from the port. A support maintenance company provides for small arms and electronics maintenance.

HARBORMASTER DETACHMENT

2-31. The harbormaster detachment employs the harbormaster command and control center to provide command and control of Army watercraft assets and provides situational awareness to maneuver commanders. The harbormaster command and control center may, as directed by the commander, be employed to supplement or expand the port manager's reach. The center can maneuver to and emplace at a geographic location that maximizes its capabilities to visually and electronically monitor the terminal area, watercraft operating within the terminal, and the waterway LOCs.

AUTOMATED CARGO DETACHMENT TEAM

2-32. The automated cargo detachment team is employed at permanent ports, inland waterways and LOTS water terminals to provide automated documentation support for cargo upload or discharge from ships. Capabilities include the ability to document breakbulk, containerized, and RO/RO cargo being discharged from up to two ships in permanent ports. Additional capabilities include: document and validate receipt of cargo for ITV and reconcile with the ship's manifest, prepare transportation control and movement documents (TCMDs) for first destination transportation, document cargo for movement by other distribution modes, and prepare discrepancy reports for the cargo accounting section. The automated cargo detachment team is normally attached to a transportation terminal battalion or DDSB. Automated cargo documentation teams use the Global Air Transportation Execution System–Surface or Transportation Coordinator's Automated Information for Movement System II to document and support ITV of bulk cargo containers, unit equipment, rolling stock, and ammunition ships moved by commercial contractors.

JOINT FORCE AND UNIFIED ACTION PARTNER ORGANIZATIONS

2-33. Joint force and unified action partner organizations provide strategic resources when conducting expeditionary intermodal operations. The following is a discussion on the major joint and unified action partner organizations that provide intermodal support during large-scale combat operations.

UNITED STATES TRANSPORTATION COMMAND

2-34. USTRANSCOM, as the DOD single manager for transportation, is responsible for providing common-user and commercial transportation, terminal management, and aerial refueling to support the global deployment, employment, sustainment, and redeployment of U.S. forces. It is also responsible for planning, allocating, routing, scheduling, and tracking assets to meet validated joint force commander deployment and distribution requirements.

AIR MOBILITY COMMAND

2-35. AMC is the Air Force component of USTRANSCOM and serves as the single port manager for air mobility. AMC aircraft provide the capability to deploy the Army anywhere in the world. AMC sources airlift capability based on priority of lift and properly sourced, verified, and validated unit line numbers via the A21 System. AMC provides both military and chartered civilian aircraft for transporting passengers and cargo, and also provides aircraft for aerial refueling operations. It also administers the Civil Reserve Air Fleet program. In this program, the DOD contracts for the services of specific aircraft owned by a U.S. entity or citizen during national emergencies and defense-oriented situations when expanded civil augmentation of military airlift activity is required. As follow-on forces to USTRANSCOM's joint task force–port opening APOD, AMC performs single port management functions necessary to support the strategic flow of the deploying forces' equipment and supplies from the APOE to the theater.

MILITARY SEALIFT COMMAND

2-36. MSC is the Navy component of USTRANSCOM. The MSC mission is to provide ocean transportation of equipment, fuel, supplies, and ammunition to sustain U.S. forces worldwide. MSC provides sealift with a fleet of both government-owned and chartered vessels. Sealift ships principally move unit equipment from the U.S. to ports in theaters of operation all over the world. In addition to sealift ships, MSC operates a fleet of prepositioned ships strategically placed around the world and loaded with equipment and supplies to sustain Army, Navy, Marine Corps, Air Force, and Defense Logistics Agency operations. These ships remain at sea ready to deploy on short notice to deliver urgently needed equipment and supplies to a theater of operations or joint operations area.

JOINT DEPLOYMENT AND DISTRIBUTION OPERATIONS CENTER

2-37. The joint deployment and distribution operations center (JDDOC) is designed to support the CCDR's operational objectives by synchronizing multimodal theater resources to maximize deployment, distribution, and sustainment capabilities. Its goal is to maximize CCDR combat effectiveness through improved total asset visibility, enabling more effective deployment and distribution.

2-38. The JDDOC is an integral component of the CCDR staff, normally under the staff supervision of the CCMD director of logistics. The JDDOC performs the following functions:

- Exercises centralized control for deployment and distribution that reliably and rapidly communicates and satisfies logistics requirements.
- Provides effective management of the transition between theater strategic and intratheater segments of the distribution system.
- Effectively links deployment and distribution process owners and other agencies to better shape support and services for military operations.
- Provides a link between the theater and the joint deployment distribution enterprise.

2-39. The TSC coordinates deployment and distribution operations with the JDDOC.

JOINT TASK FORCE – PORT OPENING

2-40. A JTF-PO is a joint expeditionary capability that enables USTRANSCOM to rapidly establish and initially operate a POD and a distribution node, facilitating port throughput in support of a contingency response. The JTF-PO is designed to use existing HN terminal infrastructure and may use support agreements and operational contracting support as required. A JTF-PO can be used when there is insufficient time to request enduring force structure through the request for forces process—such as port opening associated with a crisis action plan supporting foreign humanitarian relief or combat operations. The A/DACG is used when the request for forces process allows enough time to assemble and deploy support units. Actions taken to open and operate a POD and forward distribution node include establishing command and control, communication systems, security, and cargo and passenger handling and transfer operations.

2-41. The JTF-PO will establish communications with the appropriate CCMD and joint task force staff to tie in with the theater distribution plan but may be tasked to coordinate with the TSC for specific theater distribution requirements. Upon the arrival of a sustainment brigade with a theater opening mission, the JTF-PO should coordinate with the sustainment brigade for the disposition of cargo and passengers. The JTF-PO will usually go through the sustainment brigade for highway regulation, march credits, and coordination for additional land transportation as needed.

This page intentionally left blank.

Chapter 3

Land Terminal Operations

This chapter provides an overview of land terminal operations and the types of land terminals essential to intermodal operations. The focus of this chapter includes planning, opening, and operating the different types of land terminals.

OVERVIEW

3-1. Land terminals tie the theater strategic level to the tactical level, providing distribution to the designated point of need. Land terminals are established at interchange points along theater air, sea, inland waterway, rail, and motor transport systems. The mission of these terminals is to transship cargo and personnel carried by these modes and maintain asset visibility. Operational land terminals include APODS and SPODS, theater staging bases, theater railheads, and other theater bases and facilities. Tactical land terminals provide facilities for connecting links of the same modes when the situation dictates a change in carrier, or when force protection or manpower considerations dictate shorter line-hauls. Tactical land terminals include rail terminals, CRSPs, TTPs, inland waterways, and other interchange points between operational and tactical levels. Operational and tactical terminals support motor, rail, pipeline, and inland waterway operations.

PLANNING TERMINAL OPERATIONS

3-2. Terminals are established along air, rail, and motor mode operations networks to enable intermodal operations, which in turn enable distribution throughout the theater. Land terminals transship cargo and personnel carried by these modes. The ICTC conducts cargo transfer operations in these terminals in the theater and corps areas.

3-3. Terminal planning normally includes —
- Computing the terminal workload required. The workload is expressed as number of personnel, vehicles, containers/TEU, square feet, and STONs (for non-containerized cargo). This computation includes the total tonnage and numbers of personnel and containers that must be received, processed, and cleared through the terminal.
- Estimating the existing terminal capacity. This is the estimated total tonnage and number of personnel and containers that can be received, processed, and cleared through the terminal in a day.
- Determining repair and rehabilitation costs of existing facilities or new construction needed to increase existing terminal capacity to equal computed terminal workload (existing terminal capacity maybe insufficient to support the operational workload).
- Estimating the CHE and MHE needed to process the required workload (such as pallets, forklifts, tugs, barges, cranes) and the operators required to operate them.
- Estimating the units, personnel, civilian augmentation support, HN support, and supervisory and command requirements needed to operate the terminal.
- Coordinating force protection requirements for terminal operations.

OPERATIONAL CONSIDERATIONS FOR EXECUTING TERMINAL OPERATIONS

3-4. Physical characteristics of the terminal and layout considerations include—

- Terminal selection including HN infrastructure considerations (physical restrictions of working space and parking space that may impact capacity).
- Availability of hard surfaces in transfer areas.
- Existing facilities for storage and maintenance of MHE and other equipment.
- Proximity of transfer points to exit routes.
- Distance between loading and unloading points and temporary holding areas.
- Security and safety standoff distances.
- Berths. How many ships berths are available, and what, if any, restrictions are present?
- Available assets. What organic assets are available (jib cranes, container cranes, fork lift availability, port clearance assets, conveyors, pipelines)?
- Road networks:
  - Is there a usable road network capable of supporting port clearance operations?
  - What type of surface is there, can it be easily improved?
  - Is there a rail line available?
  - What road maintenance assets will be required?
  - What security assets will be required?
  - How vulnerable is the road network to attack or sabotage?
- Marshaling or staging areas/staging bases (See ATP 3-35 reference intermediate staging bases):
  - Are assembly or holding areas or locations on or adjacent to the port?
  - How far are they from the port?
  - Do they provide for one-way traffic into and out of the area?
  - What security assets are present?
  - What security assets need to be provided or improved?
- Characteristics of the transportation modes:
  - How many planes, trucks, or rail cars can be handled simultaneously?
  - What is the turnaround time of planes, trucks, or rail cars distributing personnel, supplies, and equipment?
  - What are the loading and unloading rates for planes, trucks, watercraft, or rail cars?
- Characteristics of cargo to be transferred:
  - Size and type of packaging. Are there any restrictions to size and type packaging?
  - Average weight of cargo units. What, if any, are the weight limitations or restrictions?
  - What types of cargo can be repackaged into smaller lots or consolidated to optimize loading?
  - Are shelter and security protective requirements available in in-transit storage areas?
  - How should fragile and perishable cargo be handled?
  - What are the requirements for handling hazardous cargo?

CENTRALIZED RECEIVING AND SHIPPING POINT

3-5.   CRSP operations established at airfields or port areas connect intermodal operations to theater opening, RSOI, and sustained distribution operations. The CRSP is an effective and efficient type of port and inland terminal operation that extends the LOCs required to provide support during deployment and RSOI operations.

3-6.   CRSPs provide centralized distribution node operations within an AO where cargo is delivered and backhaul is picked up by employing a hub and spoke concept between the CRSP and other distribution nodes. Inbound cargo arrives at an airfield or port and is off loaded, moved forward to a holding area, and transshipped to hubs such as CRSPs or forward operating bases. The intent is to facilitate throughput, reduce transloading times, maximize vehicle loads, decrease turn-around time of convoys, and reduce the number of convoys moving in the AO. As a result, cargo should flow as quickly and efficiently as possible to its destination.

3-7. Generally, cargo is not warehoused at a CRSP, with the common holding period being 24 hours or less. The objective is to receive, document, and transload cargo as quickly and efficiently as possible. Exceptions include frustrated cargo, cargo destined to low volume consignees, or battle-damaged equipment which might require inspection and processing. Under the CRSP concept, theater convoys deliver to CRSPs with CSSBs operating convoys delivering to the consignee, forward operating base, or other CRSP. Each CRSP arranges for backhaul from both the forward operating bases to the CRSP, and from the CRSP to the theater-level supply units. The overall advantage of the CRSP hub and spoke concept is that theater trucks move in and out of the AO quickly, providing faster throughput.

3-8. CRSPs also assist in container management for the theater. Materiel arriving at a theater node is sent to the CRSP distribution hub. Due to the requirement to expeditiously move materiel at a CRSP, cargo is usually containerized or palletized. As a result of cross-docking, empty commercial ocean carrier containers are identified and either made available to the ocean carrier for pickup or returned to a container holding yard. Empty government-owned containers are identified and returned to the theater base for reuse or held at the CRSP for future use. Retrograde operations can also occur at a CRSP to facilitate the disposition of materiel. See ATP 4-12 for more information on palletized cargo and container operations.

3-9. The containerized cargo operations section receives and issues customer containerized cargo, appropriately documents and tracks inbound and outbound containers, and provides a continuous and accurate on-hand status of containerized cargo.

3-10. An empty container collection point can be established to consolidate empty containers for issue to units, transport refrigerated containers to the class I yard, or retrograde detention containers. Immediately referencing Integrated Booking System Container Management Module to identify the container status (government-owned container or detention container) is a key task.

3-11. Palletized cargo operations at a CRSP involve receiving materiel in pieces or partially palletized loads. Once the materiel is received, it is consolidated by destination and shipping documents prepared for movement. Cargo can be palletized on either 463L pallets for movement on U.S. Air Force aircraft, or on wooden pallets for all other movements (truck, rail, or watercraft). Due to the requirement to palletize all cargo on U.S. Air Force aircraft, it is imperative to maintain an appropriate amount of empty 463L pallets with tie-down straps, top nets, and side nets.

3-12. Equipment and loads ready for shipment should be properly documented and placed in holding areas at the CRSP for outbound shipment to the supported units. New RFID tags are processed and attached to newly configured loads. Shipping information stored on the RFID tag is read and communicated to the national radio frequency ITV server.

3-13. Retrograde operations at the CRSP identify serviceable and non-serviceable retrograde and prepare that cargo for onward movement using government-owned containers first, and then empty carrier-owned containers, detention containers, and 463L aircraft pallets. Priority for loading containers is hazardous material and oversized items followed by all other retrograde. Maintaining separate pallets and containers for retrograde operations is recommended.

INLAND WATERWAY SYSTEM

3-14. Inland waterways include all rivers, lakes, inland channels, protected tidal waters, and canals deep enough for waterborne traffic. They include the locks, dams, bridges, and other structures that contribute to or effect movement of vessels carrying passengers and freight. ICTCs provide cargo-handling service at IWWTs. During times of peace, inland waterways are primarily used for the civilian economy. Military use depends on the degree of waterway development, necessary rehabilitation or upgrades required, the tactical situation, and the impact military use of the waterway will have on the civilian economy. It is an extremely efficient method for moving liquid, bulk, heavy, or outsized cargo where there is an abundance of navigable rivers and canals and lack of good or available roads and railroads.

3-15. The three components that makeup an IWWS are the ocean reception point (ORP), the inland waterway, and the IWWT. Engineers install the IWWS in an underdeveloped theater. In overseas theaters that have developed IWWS, the HN operates and maintains the system.

Ocean Reception Point

3-16. An ORP consists of mooring points for ships, a marshaling area for barges or other lighterage, and a control point. At least two stake barges should be at each ORP, one for import cargo and one for export cargo. Container and general cargo vessels may discharge at an ORP. Barges are then used to transship cargo to the terminal for onward movement by other modes (ground, air, or inland water). Under the stake barge system, the ORP should have water space with enough stake barges to accommodate the same number of barges as the wharf space.

3-17. The reception capacity, discharge capacity, and clearance capacity of an ORP are computed in the same way as for a marine terminal with a few minor differences. ORP clearance capacity is the number of personnel, containers, barges, or STONs of cargo that can be moved from the ORP via any mode. Terminal transfer and storage capacity influences terminal discharge capacity. Tugs and barges (terminal transfer) and wharfs or stake barges (storage) also influence ORP discharge capacity. Determining the space required and available for stake barges and the space required to move barges to and from the stake barges requires careful analysis. Transit time between the ship and the stake barge or wharf and other factors incidental to cargo, barge, or lighterage transfer and storage must also be determined.

Inland Waterway

3-18. A waterway's physical features determine its ability to carry cargo. Physical features that determine what can be moved over a waterway include—
- Restrictions of width and depth of the channel.
- Horizontal and vertical clearance of bridges.
- Number of locks, their method of operation, and the length of time required for craft to clear them.
- Freeze-ups, floods, and droughts. These also affect a waterway's capacity, and distribution planners must know when to expect these seasonal restrictions and how long they can be expected to last.
- Waterway speed, fluctuation, and direction of water current.
- Height of tidal changes.

3-19. The capacity of an inland waterway can be estimated by determining the number of craft per day that can be passed through the most limiting restriction, such as a lock, lift bridge, or narrow channel.

3-20. Turnaround time is the length of time between leaving and returning to a point. Since barges are being picked up at a wharf or stake barge, barge loading time is not part of the computation. If barges are picked up at shipside without marshaling at a wharf or stake barge, loading time of the barge would become a factor of turnaround time.

3-21. The following factors must be known to calculate turnaround time:
- Speed is influenced by the wind, current, power of craft, and size of load. If the craft's speed cannot be determined, assume it is 4 miles per hour (6.4 kilometers per hour) in still water. Speed and direction of current can frequently be discounted since resistance in one direction may be balanced by assistance in the other direction. However, this is not always the case.
- Loading and unloading time is the time to load and unload a craft at origin and destination.
- Time consumed in the locks is the time taken by a craft and its tow to pass through a lock. When exact data is lacking, lock time is assumed to be 1 hour per single lock.
- The planning factor for hours of operation per day is usually 20. Dropping barges from the tow, refueling, taking on stores, rigging up, and maintenance consume the remaining 4 hours.

3-22. Transit time is the time to move the craft the length of the haul and return to its origin. Transit time equals the distance divided by the speed of the craft. It does not include stops or delays of any kind. Due to possible damage to the inland waterway, a speed control may be in force. To determine transit time, add the following—
- The time to make up the tow.
- The distance divided by the speed of the tow.
- The time consumed passing through the locks.

● The time to break up the tow.

3-23. When determining the number of barges, tugboats, or craft required, always round up to the nearest whole number, then apply maintenance factors and round up again. See Table 3-1 for determining number of barges, tugboats, or craft.

Table 3-1. Determining factors for number of barges, tugboats, or craft

Factors	
Daily barge loading rate at ORP: A = B x C	Daily tows required at the ORP: N = F / Q
Barges loaded daily at ORP: D = A / E	Daily tows required at the IWWT: P = M / Q
Daily barge requirement at ORP: F= D + G	Turnaround time of tugboat: R = 25
Daily barge discharge rate of the IWWT: H = J x K	Number of tows a tugboat can deliver daily: T = U/R
Barges discharged daily at the IWWT: L = H/E	Number of tugboats required to deliver tows: V = (N or P*) / (T+W)
Daily barge requirement at the IWWT: M = L + G	
A – daily barge-loading rate at the ORP B – number of barge-loading berths at the ORP C – daily loading rate per barge berth at the ORP D – barges loaded daily at the ORP E – average barge cargo capacity F – daily barge requirement at the ORP G – barge maintenance factor (round up) H – daily barge discharge rate at the IWWT J – number of barge discharge berths at the IWWT K – daily discharge rate per barge berth at the IWWT L – barges discharged daily at the IWWT	M – daily barge requirement at the IWWT N – daily tows required at the ORP P – daily tows required at the IWWT Q – barges per tow R – turnaround time of a tugboat in hours S – transit time for tugboat T – number of tows a tugboat can deliver daily U – operational hours per day V – number of tugboats required to deliver tows W – tugboat maintenance factor (round up) *Largest of the two (N or P)
IWWT inland waterway terminal ORP ocean reception point	

Inland Waterway Terminal

3-24. An IWWT normally includes facilities for mooring, cargo loading and unloading, dispatch and control, and repair and service of all craft that can navigate the waterway. Terminals either exist or are established at the origin and terminus of the inland water route. Intermediate terminals are located along the way, wherever a change in transportation mode is required. Terminals on an IWWS can be classified as general cargo, container, liquid, or dry bulk commodity shipping points. Terminals of the three latter types usually include special loading and discharge equipment that permits rapid handling of great volumes of cargo.

3-25. The number of cargo transfer units required to support IWWTs depends on the results of an IWWT throughput analysis. An analysis is conducted for each IWWT in the IWWS. The combined capacity of the IWWTs is the cumulative total of the restricting capacity (reception capacity, discharge capacity, or clearance capacity) for each IWWT. There may be a requirement for tugboats stationed at the IWWTs to make up or breakup tows and shift barges between terminals and an additional mooring area. The additional mooring area may be required to allow a buildup of barges to keep an even flow of barges at the terminals.

3-26. After estimating the capacity of the three functional components of the IWWS, the least of the three capacities is used as the estimated capacity for the entire system (see Table 3-2). Once the capacity of the IWWS has been determined, cargo transfer requirements for each component of the IWWS can be determined.

Table 3-2. Daily IWWS capacity

Ocean Reception Point	Inland Waterway	Inland Waterway Terminal
3,000 tons	2,000 tons	2,500 tons

MOTOR TRANSPORT TERMINALS

3-27. Motor transport terminals are normally located at both ends of a line-haul operation. They form the connecting link between local hauls and the line-haul service. They may also be located at intermediate points along the line-haul route where terrain necessitates a change in type of carrier. Cargo transfer elements provide cargo-handling service at motor transport terminals. See ATP 4-11 for more information.

TRAILER TRANSFER POINTS

3-28. TTPs are established along the line-haul system to divide the movement into legs. At TTPs, semitrailers or flatracks (intermodal platforms) are exchanged between line haul vehicles operating over adjoining legs of a line haul route. TTP functions also include reporting, vehicle and cargo inspections, documentation, and dispatching. TTPs are often co-located at hubs or operating bases.

3-29. Line-haul tractors arriving from rear areas deliver loaded semitrailers at TTPs and pick up empty or retrograde semitrailers for return movement. Line haul tractors coming in from forward areas drop their empty or return-loaded semitrailers and pick up the forward-moving loads for further movement toward ultimate destinations. Shuttle tractors may be used within the TTP to spot and prepare semitrailers for movement. This action reduces turnaround time of line haul tractors and makes the operation more efficient. The distance of a line haul leg is based on a 10-hour shift per driver and 1 hour of delay. Therefore, the optimum one-way travel time between TTPs is 4.5 hours. Using this planning factor, each driver can complete one round trip per shift. This eliminates the need for billeting drivers away from their assigned unit, provides rested drivers for each trip, and allows for vehicle maintenance.

3-30. The following are the general responsibilities of the TTP team:
- Provides a central headquarters for all movement regulating points.
- Normally operates at echelons above brigade but may be required to operate in a brigade area.
- Receives, segregates, assembles, and dispatches up to 250 loaded or empty semitrailers and 125 tractors per day.
- Provides emergency refueling and repair of vehicles transiting the TTP.
- Provides inspection and emergency repairs of up to 10 percent of transit vehicles and semitrailers with organic mechanics.
- Provides area recovery of disabled vehicles operating in the line haul operation.
- Prepares and maintains operational records and reports on organic equipment.
- Assists in the coordinated defense of the unit's area or installation.
- Performs field-level maintenance on wheeled vehicles.
- Coordinates with higher and adjacent headquarters for windows of opportunity to move and reconstitute operations rapidly in another location.
- Plans for rapid movement and setup to enable TTP operations in alternate locations.

3-31. Figure 3-1 depicts a notional TTP layout.

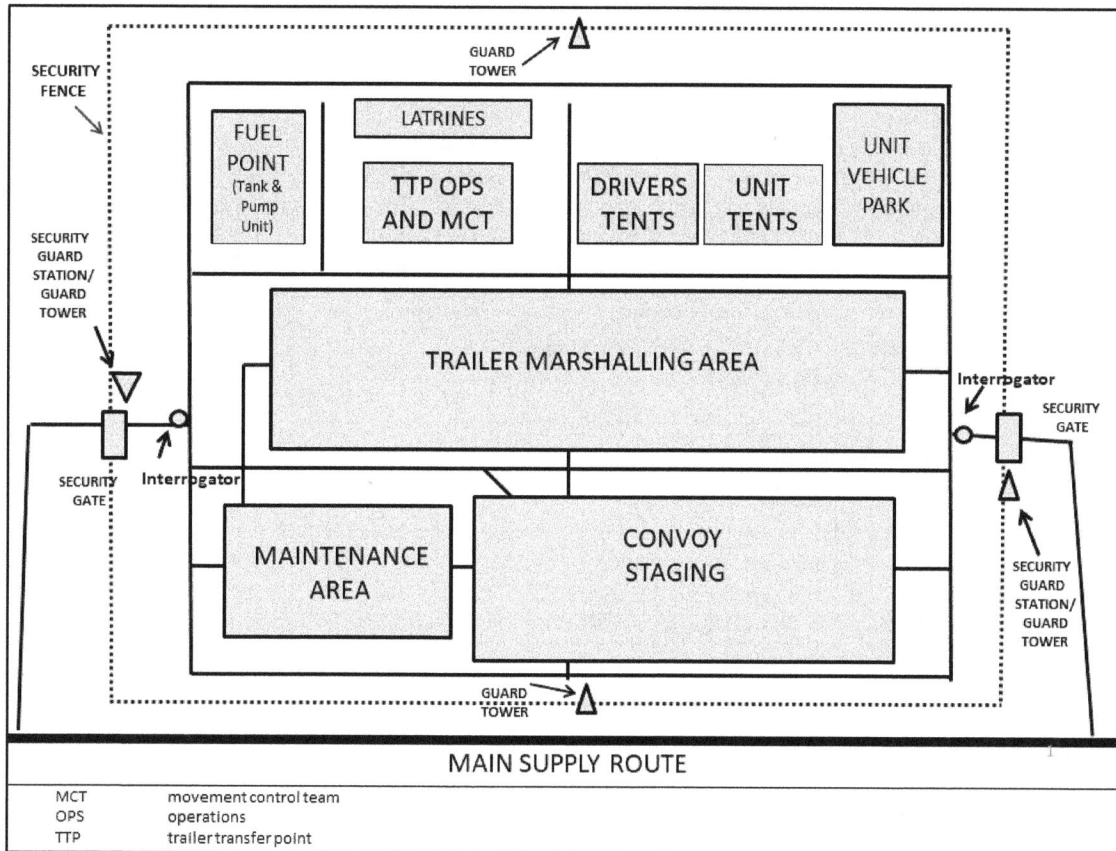

Figure 3-1. Notional trailer transfer point

RAIL TERMINALS

3-32. The Army does not own or operate a rail capability when operating OCONUS. All rail support should be coordinated with HN authorities to schedule movement. When sufficient rail infrastructure exists for movement of cargo and personnel in a theater, it can enable the RSOI process. Rail is usually the preferred mode to move tracked vehicles and heavy equipment to ports of embarkation or debarkation. Although the Army does not own or operate rail cars, it can support rail operations by providing MCTs to coordinate movement and ITV of cargo. The Army can also provide ICTCs capable of staging and marshaling equipment and assisting in the upload and download of equipment from rail cars. The Army also has railway advisors within the expeditionary railway center (ERC) to assist with planning and rail operations.

3-33. Planning to use rail assets begins with determining which type of rail system (cargo, passenger, or combination) will be utilized, which in turn determines the terminal type. Rail plans should be integrated into the overall movements plan for a theater. Considerations for rail terminal planning include cargo type, passenger number requirements, and estimated capability throughput. During initial planning, rail operations usually consist of a pre-invasion plan based on available information and intelligence. This initial information and intelligence provide general estimates of potential rail terminal movement capabilities in theater.

3-34. Terminal types can include rail yards, freight stations, or passenger stations. Rail plans should be integrated into the overall movements plan for a theater. Rail is primarily a theater strategic and operational level of war asset. Rail plans can be integrated in a CCMDs war plans. When sufficient rail infrastructure exists for movement of freight and personnel in a theater, it can enable RSOI and opening the theater processes. Rail is the preferred mode to move bulk freight, which includes items like petroleum and tracked vehicles, to ports of embarkation or debarkation. Elements that can support expeditionary railway operations include MCTs and ICTCs that are capable of staging and marshaling equipment; petroleum units; and capabilities to move liquid freight, upload and download rail cars, and facilitate onward movement of passengers and equipment. HN support is needed to support these operations in most situations. See ATP 4-14 and ATP 4-43 for additional planning considerations and operations for rail terminals.

EXPEDITIONARY RAILWAY CENTER

3-35. Army rail forces have shifted to planning, advisory, capability assessment, and coordinating roles and have relinquished the role of providing operating control over HN rail capability. Army units designed to primarily support rail operations are in the Army Reserve. The ERC is a force structure within the Army Reserve that consists of Army rail experts that perform six key functions:

- Provide rail network capability and infrastructure assessments.
- Perform rail mode feasibility studies and advise on employment of rail capabilities.
- Coordinate rail and bridge safety assessments.
- Perform and assist with rail planning.
- Coordinate use of HN or contracted rail assets.
- Perform contracting officer's representative duties to oversee contracts and provide quality assurance of the contracts.

3-36. The focus of the ERC is planning and coordinating rail operations within a theater of operations. The ERC can also focus on enhancing strategic and operational throughput (such as port clearance via rail) and provide contracting officer's representative oversight. The ERC provides the required rail expertise to accomplish all of this to the CCDR; the TSC, ESC, and their subordinate sustainment brigades; and the HN.

To meet these requirements, the Army redesigned the rail unit structure and formed a single ERC consisting of a headquarters and five deployable railway planning and advisory teams, all totaling 181 personnel. The mission of this organization is not to operate a railroad, but rather to perform capability assessments, serve as the combatant and/or sustainment commander's adviser, and advise and assist HN and contracted rail personnel. The successful use of the rail mode in the conduct of intermodal operations, provided by the oversight of the ERC and its subordinate teams, has the potential to increase operational movement capability by millions of STONs and decrease the logistics footprint. The modular nature of the organization and its ability to plug into multiple levels of command across the range of operations will enhance this capability. See ATP 4-14 for more information.

Chapter 4

Air Terminal Operations

This chapter provides an overview of Army air terminal operations, airfield organization, and discusses the A/DACG and its support of airfield operations.

OVERVIEW

4-1.   Air terminals are airfields with aerial port facilities capable of accepting, processing, and manifesting passengers and cargo. Air terminal intermodal capabilities enable transshipment and transfer of cargo and passengers for onward movement to their final destination. Air terminals support the deployment, reception, and onward movement of the forces, cargo, and supplies required to sustain large-scale combat operations. For more information see ATP 3-35.

AIR TERMINAL OPERATIONS

4-2.   An air terminal may be a military airfield or civilian airport. The airfield and the entire system of supporting facilities required to handle inbound and outbound passengers and cargo is collectively known as a joint aerial port complex. The joint aerial port complex containing an air terminal is a key node in any deployment, redeployment, or sustainment operation. During these operations, the aerial port complex handles flow in both directions, operating as both an APOE and APOD. This includes the reception of unit personnel and equipment, replacement personnel, and sustainment cargo, as well as the retrograde movement of noncombatants, wounded personnel, enemy prisoners of war, human remains, and equipment requiring repair.

4-3.   Air terminal operations begin once the A/DACG is in place and synchronized with its air mobility force counterparts. An ICTC can provide personnel and equipment to load and off-load aircraft, transport cargo for airfield clearance, and process outbound equipment. Air terminal operations can be separated into two functions—arrival and departure operations (see Figure 4-1 on page 4-2). Regardless of whether an airfield is designated as an APOE or APOD, the process required to receive and dispatch passengers, equipment, and supplies remains the same. Airfields designated as either an APOE or APOD may have a larger throughput capacity and will require additional resources compared to other air terminals in the AO.

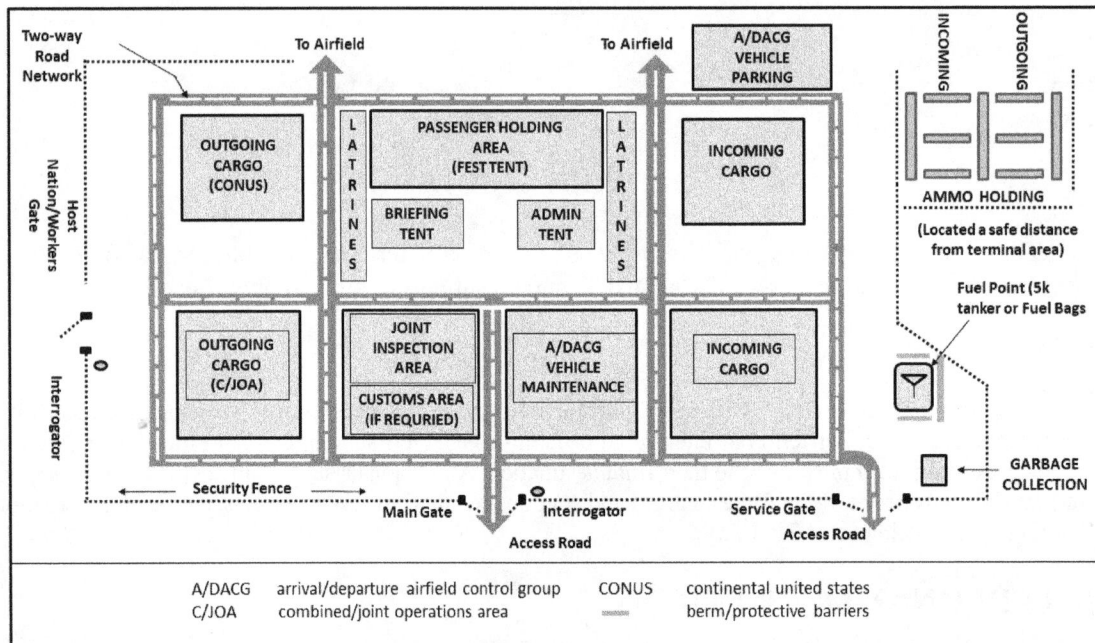

Figure 4-1. Notional air terminal layout

4-4.   Air terminal intermodal services require proper MHE and CHE, pallets, and cargo documentation capabilities to unload, segregate, and load conveyances in order to transship cargo whenever a change in mode occurs.

AIR TERMINAL ORGANIZATION

4-5.   The departure airfield is usually organized around four separate activities: marshaling area, alert holding area, call forward area, and the ready line and loading ramp area. These areas may or may not be adjoining.

4-6.   The marshaling area is where units move to complete vehicle and cargo preparations for aircraft loading for redeployment or intratheater movement. The marshaling area can include temporary, fixed, or field facilities for transportation, communication, and lodging and the areas that support those functions. The unit is responsible for activities conducted within the marshaling area. In this area, the unit prepares for air movement by assembling vehicles, equipment, supplies, and personnel into chalks.

4-7.   The alert holding area is the control area for vehicles, equipment, cargo, and personnel. It is used to assemble, inspect, hold, and secure aircraft loads; this area is also referred to as the outbound yard. The A/DACG is responsible for activities conducted within the holding area. The A/DACG assumes control of the chalks once the unit has completed its initial preparations for movement and after the A/DACG issues a call for movement from the holding area to the call forward area. If the air mobility force has the responsibility for passenger manifesting, the A/DACG will not be required to establish a holding area for passengers.

4-8.   The A/DACG is responsible for activities conducted within the call forward area and can receive assistance from the air mobility force. In this area, the unit, A/DACG, and air mobility force members conduct a joint inspection and correct discrepancies. This is the final check to ensure that all cargo and equipment is properly prepared and documented for safe and efficient air shipment.

4-9.   The air mobility force controls activities conducted within the ready line and loading ramp area. In this area, the air mobility force receives vehicles, equipment, cargo, and personnel from the call forward area; directs aircraft loading in conjunction with aircraft load masters; supervises the supported Service while loading and restraining cargo aboard aircraft; conducts additional briefings; and performs inspections, as required, to facilitate loading of the aircraft. The A/DACG may perform this function at some air terminals if the air mobility force capability is not present.

4-10. The A/DACG coordinates with the deploying unit and arriving passengers to provide off-load teams for baggage pallets and vehicles. The A/DACG should establish procedures to retain accountability of pallets, nets, and shipping containers throughout the reception process and return shipping equipment to the air mobility force for retrograde as soon as practical.

4-11. In the alert holding area, the A/DACG assembles personnel, cargo, and equipment for movement to unit marshaling areas and maintains accountability of cargo until local unit pickup or until requested transportation arrives. The A/DACG will inbound clear departing cargo to the destination MCT when it is not a unit pickup. Other holding area tasks include—

- Maintaining and reporting cargo and passenger arrivals in the appropriate information automation system.
- Collecting and returning all aircraft pallets, nets, shipping containers, and dunnage to the air mobility force.
- Accepting, inventorying, and controlling the aircraft loads.
- Establishing a discrepancy correction area for cargo and documentation.
- Inspecting documentation for accuracy and completeness.
- Ensuring passengers are accounted for.
- Providing emergency maintenance, petroleum, oil, and lubricants (including defueling capability), and related services.
- Establishing a traffic flow pattern.

4-12. Unit marshaling or forward node areas should be established to allow for rapid terminal clearance, reducing port congestion and the potential for slowdowns or work stoppages in off-loading operations. These areas could be located on the same base as the airfield or on another base in close proximity to free up the limited staging space at air terminals. The A/DACG should have organic trucks to assist in moving cargo from the terminal, but additional truck support is needed for terminal clearance to support reception and onward movement or increased sustainment cargo flows.

4-13. A departure operation at an air terminal is the process of preparing and loading personnel and equipment for air movement from the terminal. The process includes manifesting and air load planning, preparing and inspecting equipment, and loading the aircraft. As with the debarkation process, the tasks for the A/DACG will vary depending on the air mobility force capabilities. Below are some additional A/DACG functions:

- Coordinate and establish a passenger processing or holding area.
- Determine passenger eligibility.
- Brief passengers on departure times.
- Weigh troops and baggage.
- Manifest passengers or coordinate manifesting procedures.
- Escort passengers to and from the aircraft.

ARRIVAL/DEPARTURE AIRFIELD CONTROL GROUP

4-14. The A/DACG is an ad hoc organization established to control and support the arrival and departure of personnel, equipment, and sustainment cargo at airfields (see Figure 4-2 on page 4-5). Essential tasks to open and operate the APOD and forward distribution node include—

- Establishing command and control.
- Ensuring communication systems are functional.
- Planning for and providing area security.
- Conducting cargo and passenger transfer operations.
- Confirming cargo documentation procedures.
- Maintaining ITV and RFID networks.

4-15. The A/DACG coordinates with Air Force contingency response group/contingency response elements to establish communications, passenger and cargo discharge procedures, and locations for the off-load ramp and cargo holding areas, and to set a general land usage arrangement for the aerial port layout.

4-16. The A/DACG also coordinates with the deploying unit and arriving passengers to provide off-load teams for baggage pallets and vehicles. The A/DACG establishes procedures to ensure that accountability of pallets, nets, and shipping containers is retained throughout the reception process, and that shipping equipment is returned to the air mobility force for retrograde as soon as practical.

4-17. Elements of an MCT and an ICTC typically operate the A/DACG. However, the mission can be performed by any unit with properly trained personnel and the appropriate equipment. An MCT acts as the Army liaison to the Air Force and can provide a detachment-level command and control structure, passenger processing, air load planning, loading coordination, cargo documentation, and onward movement for cargo and passengers. Elements of an ICTC can augment an MCT to provide the personnel and equipment required to load and off-load aircraft as needed, transport cargo for airfield clearance, process outbound equipment, and provide minor maintenance support.

4-18. An arrival operation at an air terminal is the process of receiving passengers and cargo via airlift. Numerous functions occur during this process to include—

- Off-loading cargo, equipment and supplies.
- Movement to the marshaling area.
- Providing essential field services.
- Clearing personnel, equipment, and cargo from the terminal.
- Maintaining ITV and radio frequency networks.

4-19. The main areas of the airfield for debarkation are the off-loading ramp, holding area, and unit marshaling area. The A/DACG and air mobility force will ensure that arriving aircraft are off-loaded in a timely manner and that equipment, supplies, and personnel proceed immediately to the holding area. Normally the air mobility force is responsible for the unloading of aircraft, but the A/DACG could be responsible for the physical off-load of the aircraft for small airfields or airfields supporting a brigade combat team or smaller size unit. The air mobility force is also normally responsible for passenger reception, but the A/DACG can meet arriving passengers at the aircraft and process them through the terminal if the air mobility force capability is not available.

4-20. Off-load ramp activities are controlled by the air mobility force. Each load will be released to the A/DACG for return to unit control at the holding area, sometimes referred to as the inbound yard. Arrival of personnel and equipment can coincide with arrival or draw of equipment, either at the APOD, SPOD, or prepositioned stock sites. When unit personnel arrive, they may move—

- Directly to a unit marshaling area if the unit moves with its equipment.
- To prepositioned stock sites to receive equipment.
- To aircraft for intratheater air movement (air-to-air interface).
- To the SPOD to receive unit equipment off-loaded from ships.
- To holding areas if equipment arrival is delayed.

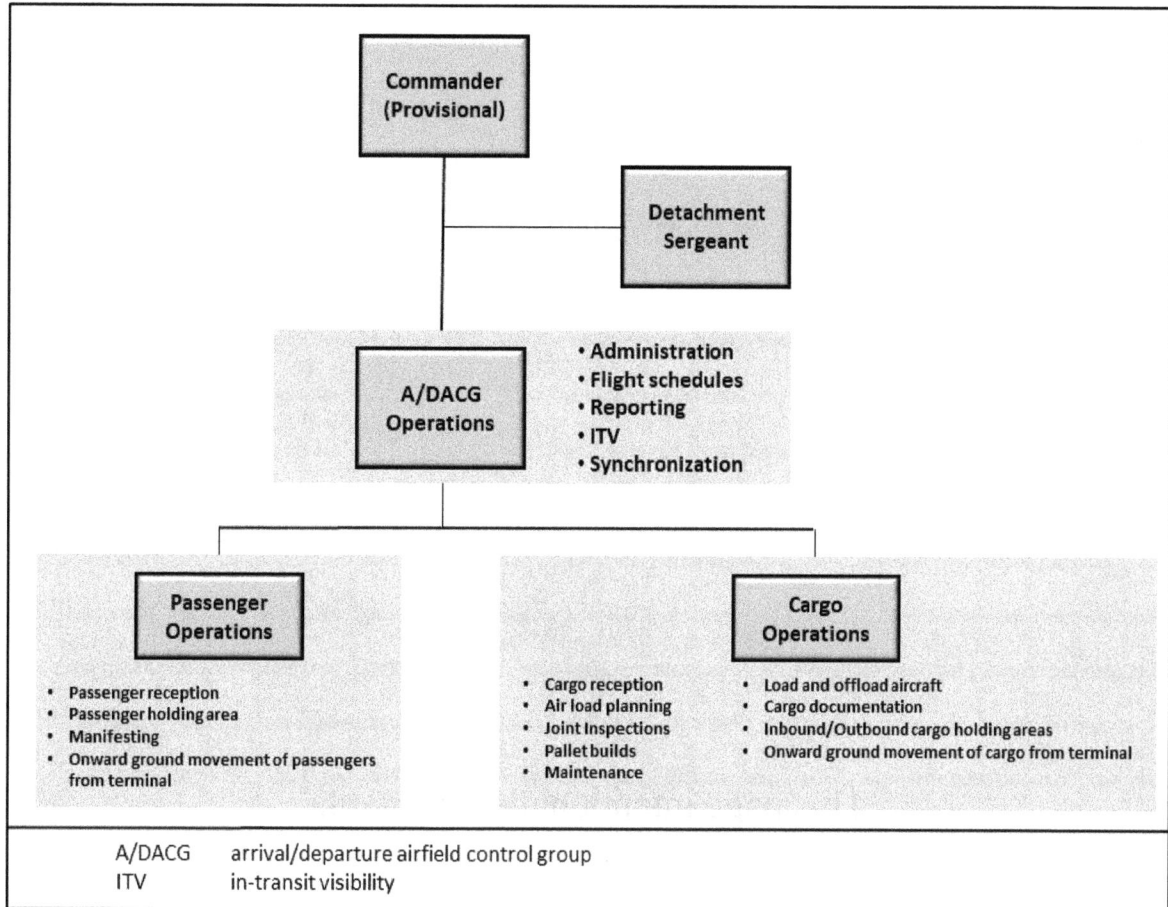

Figure 4-2. Notional A/DACG structure

This page intentionally left blank.

Chapter 5

Maritime Terminal Operations

This chapter will discuss the fundamentals of planning, opening, and operating maritime terminals at permanent ports, unimproved ports, and at offshore locations.

OVERVIEW

5-1. Maritime terminal operations include loading and unloading cargo from various types of ships and watercraft for delivery, transshipment, and onward movement of cargo and personnel. Maritime terminal operations are conducted at permanent or developed ports, unimproved or degraded ports, bare beaches, and at offshore anchorages.

MARITIME TERMINALS

5-2. Water terminal operations are conducted at ports, bare beaches, or offshore anchorages. Terminals will vary in size, function, and capability; each of these factors are considered during port operations planning.

PERMANENT PORT FACILITIES

5-3. Permanent port facilities generally contain sophisticated cargo handling systems designed for transferring oceangoing freight; these systems enable vessel loading and discharge operations, reception and staging operations, and port clearance. These facilities have sufficient water depth and pier length to accommodate deep-draft vessels. Typically, at permanent port facilities, deep-draft oceangoing vessels come alongside a pier or quay and discharge cargo directly onto the apron of the wharf or pier. Most cargo either moves into in-transit storage facilities to await terminal clearance or is discharged directly to land transport. Permanent port facilities are the most capable terminals for military operations such as RSOI and large-scale combat operations, which require handling large volumes of equipment and containerized cargo. A permanent port facility operated by an HN under contract requires operational contract support. See JP 4-10 and ATP 4-10 for more information on operational contract support.

UNIMPROVED PORT FACILITIES

5-4. Unimproved port facilities are not usually designed to handle the same volume of cargo as improved port facilities. These terminals can include naturally austere or damaged ports where the cargo handling capacity is limited due to a reduction of port infrastructure. Characteristics of unimproved facilities include reduced sources of labor, and insufficient water depth and pier length to accommodate oceangoing cargo vessels. Therefore, use of shallow-draft lighterage can enable discharging ocean-going vessels that are anchored in stream or off-shore. This method of handling cargo involves using self-propelled watercraft to carry cargo between a ship and an unimproved port facility. When Army units are the only Service involved, it is a LOTS operation (see chapter 6). In most instances, Army cargo transfer units employ organic assets to operate an unimproved port facility. Operations at this type of facility are established when permanent maritime terminals are not available or an alternate discharge location is desired.

BARE BEACH

5-5. Bare beach locations usually lack organic equipment and any significant port infrastructure capabilities required to conduct cargo discharge operations. Due to the diminished capability at these locations, LOTS provides the best means to conduct port operations. Bare beach facilities should be established when no other terminal facilities are available, additional throughput is required, or as an alternate discharge site. When a bare beach is the only option available for transferring cargo from strategic sealift assets to the theater or AO,

Army personnel and equipment can be brought in to set up capabilities to facilitate offload of vessels anchored at sea. Capabilities brought in can include—

- Floating causeway piers (sometimes configured as trident piers).
- Matting surfaces deployed across beaches to support movement of tracked vehicles.
- CHE to support lifting of containers.

5-6. LOTS operations provide a critical capability for bringing equipment, cargo, supplies, ammunition, and petroleum products into theater because of degraded or austere port facilities, or they can be used to bypass enemy anti-access or area denial efforts. LOTS operations can also be performed for bulk liquid supplies such as fuel and water. LOTS operations use pipelines and hose lines to offload tankers at undeveloped ports into tactical petroleum terminals operated by the petroleum pipeline terminal operating company. The petroleum support battalion directs the operation of petroleum port terminal facilities and storage facilities. Transportation medium truck companies transport bulk petroleum inland. See ATP 4-43 for information on Army petroleum operations.

5-7. Army watercraft such as landing craft and causeway system lighterage will be employed to move cargo from sealift ships to the beach or causeway piers for discharge and RSOI operations. Beach terminals require select sites to support delivery of cargo by lighterage across the beach and movement into marshaling yards or onto waiting clearance transportation.

WATER PORT OPENING

5-8. Water port opening includes the activities required to establish, initially operate, receive deploying forces (equipment and cargo), and facilitate throughput at SPODs. Initial port opening capabilities should be in place in advance of deployment forces, sustainment, or humanitarian relief supplies. The TSC and its subordinate sustainment brigades, terminal battalions, and SOCs perform the port operator functions at SPODs.

5-9. Water port tasks executed by TSC and its subordinate elements include—

- Port preparations and improvement.
- Cargo discharge and upload operations.
- Harbor craft services.
- Port clearance and cargo documentation activities.

5-10. SDDC is the single port manager for all common-user SPODs. It recommends ports to meet operational requirements and is primarily responsible for planning, organizing, and directing the operations at the seaport.

5-11. Water port management tasks executed by SDDC include—

- Perform port, transportation network, and forward distribution node assessment and surveys.
- Open the SPOD and begin seaport clearance operations.
- Prepare for SPOD management operations.
- Establish ITV and RFID networks on the port and at the forward distribution node.
- Initiate intermodal platform management procedures.
- Establish initial movement control capability if necessary.
- Establish staging areas.
- Facilitate the unit's ability to receive, re-assemble, and organize the cargo for operations.
- Manage the port support activity for discharge operations as required.
- Facilitate and enable the CCDR's joint RSOI operations.
- Establish and provide interface for operational and HN contracting.

PLANNING MARITIME TERMINAL OPERATIONS

5-12. Terminals facilitate movement of cargo in and out of the port. Incoming ships are directed to specified terminals for discharge based on the following factors:

- Availability of assets at the terminal to support discharge operations.

- Temporary storage locations for discharged cargo.
- Capabilities of the distribution network.

5-13. Maritime terminals present security and force protection challenges not faced at other terminals. Force protection measures should include consideration for vessels both in port and moored offshore. U.S. Coast Guard elements and divers may be required to ensure adequate protection of the vessels in port and moored offshore. Roving patrols and gate protection should be considered to counter threats originating from outside of the port. Security checks should be established pier side to guard against threats to vessels during loading operations. Figure 5-1 depicts a notional water terminal layout.

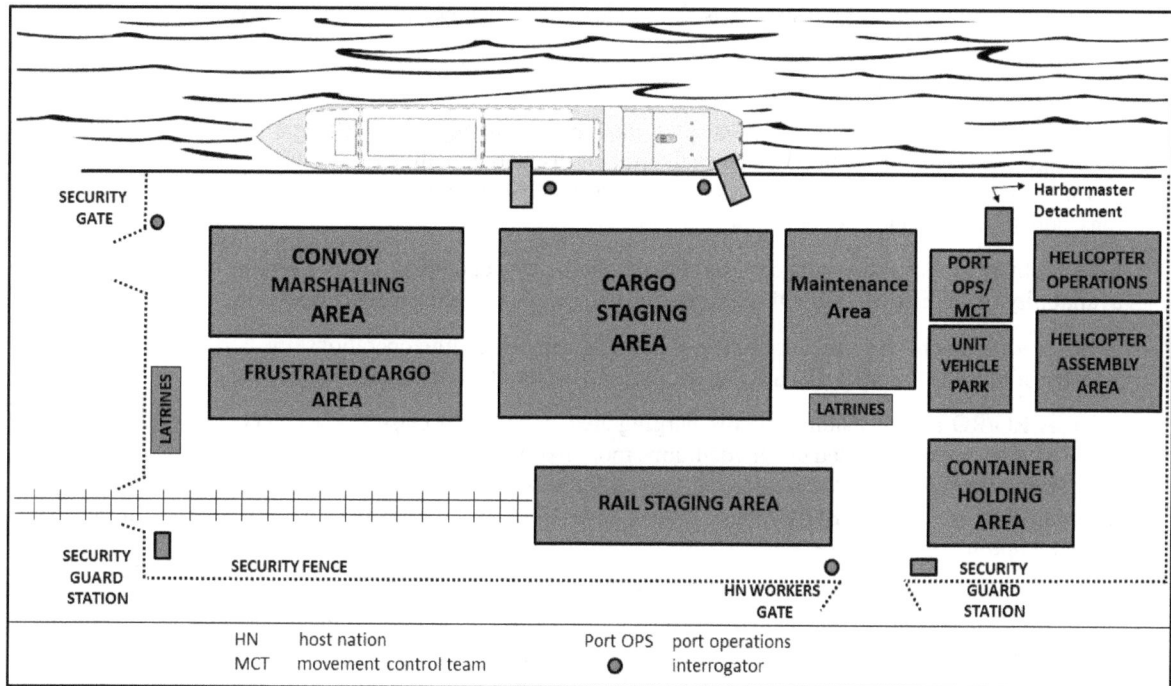

Figure 5-1. Notional water terminal

5-14. Vessel arrival at the port involves berthing, port operations, and customs clearance IAW Defense Transportation Regulation (DTR) 4500.9-R, Part V. A boarding party boards the ship to coordinate with the vessel master before moving or unloading cargo. The boarding party may consist of only the boarding officer, which is normally the terminal battalion operations officer or SOC commander.

5-15. During this visit and inspection of the ship and cargo, the boarding party may decide to alter the discharge plan that was made before the ship arrived. Unforeseen conditions such as damage to the ship's gear, unexpected priority cargo, or oversize or heavy lifts not noted on advanced stow plans may change the initial discharge plan. Specific tasks performed during inspections include—

- Determine and report the general condition of the ship's equipment and facilities.
- Communicate pertinent information to the vessel master and the military troop commander.
- Determine major damage to or pilferage of cargo and obtain other information pertinent to unloading the vessel's cargo.
- Customs representatives check for clearances, narcotics, weapons, and contraband, and perform other necessary customs activities according to theater directives and HN laws.
- MSC representative determines from the ship's officers the requirements for repairs, fuel, and storage and delivers MSC instructions to the vessel master.
- Medical personnel check for communicable diseases, sanitary conditions of personnel spaces and facilities, and condition of perishable cargo.
- Harbormaster coordinates berthing, tug assistance, and employment of floating cranes and other harbor craft assigned.

- Coordinate discharge plan with the vessel master.
- Coordinate plans for using watercraft to unload vessels at anchorage berths.
- Coordinate plans for personnel movements through the terminal.

WATER TERMINAL OPERATIONS

5-16. Water terminal operations are conducted at a port, bare beach, or offshore anchorage. Terminal operations include planning and conducting activities to load and unload vessels, whether they are Army watercraft (organic), MSC ships, or commercial ships. Key water terminal activities include discharge operations, cargo disposition instructions, cargo clearance, documentation, and marshaling.

DISCHARGE OPERATIONS

5-17. The cumulative amount of cargo that can be discharged from each type of berth is the terminal discharge capacity. This is based on an evaluation of discharge facilities and equipment found on the berths as well as on the type of ship to be docked on the berths. It is expressed in STONs, containers, measurement tons, square feet, or numbers of personnel.

5-18. Estimated discharge capacity for breakbulk berths operating on a 24-hour basis at 75 percent availability of CHE is 1,875 STONs of breakbulk cargo discharged each day per berth.

5-19. For lighters berth using one crane per lighter during discharge operations, the berth can discharge 300 STONs of breakbulk cargo, 450 STONs of ammunition, or 200 containers per day.

5-20. A RO/RO berth's loading and discharging areas for various classes of RO/RO vessels varies greatly. Since MSC vessels are loaded under conditions more likely to be encountered during a military contingency, their short-term rate of 600 measurement tons or 3,898 square feet of cargo per hour is recommended for planning purposes. A RO/RO terminal should have 10 acres of open hard surface space with at least a 100-foot apron.

5-21. Underdeveloped container berths have a discharge rate of 300 containers per day that applies when off-loading or loading containers using heavy lift cranes working at anchor alongside a ship in an underdeveloped permanent port. If back-loading is conducted at the same time as off-loading, the back-loading rate equals about one-half of the discharge rate for off-loading only. This berth should have at least a 100-foot apron.

5-22. Developed permanent container terminals using container-handling cranes have a simultaneous discharge and loading rate between 700 and 800 containers per 24-hour period. The rate of discharge at any container terminal depends on the type of CHE, type of ship being worked, and the number of container cranes used. The number of cranes per terminal and berth often varies between terminals. The size of the container does not affect the rate of discharge. If container handling and transport equipment is available, all containers can be handled at the same rate. Also, barges have a set discharge rate at a developed permanent container terminal of two barges every 25 minutes and one container every 3 minutes (if containers are carried in lieu of barges on the main deck).

5-23. Many factors affect production during discharge operations. Weather, sea state, visibility (fog and darkness), crew experience, the type of lifting gear (shore crane or ship's gear), cargo stow, tactical situation, and terminal congestion and packaging all affect discharge production. The sum of these positive and negative influences results in the number of lifts that can be obtained per hour. Lift capacity can be computed by hatch or for the entire vessel. It can be obtained by timing the lifts for a specified period or by computing information from tally sheets at the end of a shift.

CARGO DISPOSITION

5-24. Cargo disposition instructions are based upon cargo destination information and mode of transport to clear the cargo from the terminal. The TSC issues cargo disposition instructions and determines the mode of transport required to move cargo from the terminal of discharge to its destination. Cargo disposition instructions are used as an advance notification document for consignees of cargo. This information, along with vessel manifest information, is relayed to the terminal battalion responsible for the terminal where the vessel is to be discharged. An MCT coordinates mode of transportation to move the cargo and coordinates with the MCT located at the consignee destination location to schedule delivery.

5-25. After the disposition of the incoming ship is decided, the terminal brigade must coordinate a number of actions before ship discharge and port clearance operations can commence. These actions mainly consist of the following:

- Receive detailed disposition instructions for military and civilian air cargo, including diversions and detailed routing instructions from the TSC.
- Arrange clearance of personnel and cargo forward, bypassing rear area facilities.
- The TSC ensures MCTs are available at each discharge site to assist terminal personnel.

CARGO CLEARANCE

5-26. Cargo clearance is the function of moving cargo from shipside or temporary storage to its first destination outside the terminal operating area. This first destination may be the final destination, or it may be a rear area depot. An MCT coordinates cargo clearance from the terminal. The following conditions may impact cargo clearance:

- Lack of proper CHE or MHE.
- Lack of proper transport.
- Inability of receiving locations to accept cargo.
- Delays in receiving cargo disposition instructions.

5-27. Temporary in-transit storage areas are employed if cargo can't be cleared from the terminal. If temporary holding is necessary, the cargo held should not exceed one day's discharge. The areas should have a hard, all-weather surface and should be located between the discharge points and the inland transportation loading area. This would permit efficient use of MHE to move cargo from shipside to the area, within the area, and from the area to the transportation loading area. Emergency supplies and equipment for containing hazardous material spills should be readily available at or near temporary storage areas.

MARSHALING

5-28. Marshaling is a rapid and efficient means to control movement and track discharged cargo between ship and shore. Cargo marshaling yards provide a location to hold and process cargo pending further movement. Intermodal capabilities should be available to promote efficient processing and documenting of transshipments for onward movement. See appendix A for marshaling yard operations.

DOCUMENTATION

5-29. Cargo moving through Army terminals is documented according to DTR 4500.9-R, Part II. The basic document for all cargo movements under these procedures is DD Form 1384 (*Transportation Control and Movement Document*).

DD Form 1384

5-30. DD Form 1384 is a multipurpose form that can be prepared in a manual or automated format. The manual version of the form is a seven-part document. Originated by the shipper for each transportation unit, the TCMD data (not necessarily the document) accompanies the shipment from the origin to the consignee. Detailed procedures for preparing and processing the TCMD and multinational documents are in DTR 4500.9-R, Part II. The TCMD is used—

- To provide advance notice of shipment to consignees.
- As an air bill, a highway waybill, a dock receipt, and a cargo delivery report.
- For movement control of shipments worldwide within the DOD transportation system, including in-transit reporting and tracing actions.
- As a source document for mechanically prepared air and ocean manifests.
- As a source of logistic management data.
- As basis to prepare bills of lading, freight warrants, and train manifests as required.

5-31. The TCMD is normally the basic document for checking and documenting incoming cargo. However, other forms such as tally sheets may be used for internal accountability. When drafts of cargo are moved away from the ship, the cargo checkers begin internal accountability. Throughout the terminal, cargo checkers check the cargo in and out and direct cargo to its next destination. The TCMD will be properly annotated when cargo is put into the in-transit storage area or loaded aboard the clearance conveyance. The unit commander is responsible for the checkers and determines how often the cargo must be checked. The system must be sound and must allow a smooth and constant flow of the cargo with accurate accountability.

5-32. Except when cargo is moved directly from shipside to a local consignee, cargo must be reconfigured into transportation units, such as railcar loads or line-haul truckloads, before clearing the terminal area. These units may differ from those in which the cargo left shipside and may require new TCMDs. Copies of these new and more complete TCMDs accompany the cargo to the destination.

Daily Operations Report

5-33. In addition to the documentation required by existing regulations, SDDC or the sustainment brigade will normally require each terminal battalion operating a port or beach terminal to prepare a daily operations report. This report usually includes the following:

- Number of passengers embarked, debarked, and awaiting embarkation and debarkation, and the number of passengers to be handled during the next 24 hours.
- Number of tons (weight and measurement ton) of cargo by major category (general, vehicles, and petroleum, oil, and lubricants) that have been discharged, loaded, and cleared (by mode); awaiting discharge, loading, and clearance; and the number of tons booked and expected in the next 24 hours.
- Number of ships which have arrived, departed, remain in port, and are expected to arrive and depart during the next 24 hours, and the status of ships in port (for example, discharging, loading, awaiting orders, and under repair).
- Workload for the previous months and anticipated workload for the next month.
- Summaries of available berths, number and capacity of lighters and trucks, number of gangs for ship and pier work, covered and open storage space, number of railroad cars that can be accommodated and cleared, and MHE or CHE availability.

Chapter 6

Logistics Over-the-Shore Operations

Chapter 6 discusses LOTS as a unique intermodal transportation capability designed to support deployment, sustainment, and redeployment operations in a compromised environment. This chapter will focus on the procedures for planning, opening and operating a LOTS operation.

OVERVIEW

6-1. LOTS operations enable the loading and off-loading of ships in austere areas where permanent port facilities are damaged, unavailable, or inadequate for operational needs. The TBX is normally tasked to execute LOTS operations with augmentation from other support elements including engineers, petroleum units, and MCTs.

PLANNING FOR LOTS OPERATIONS

6-2. LOTS operations are conducted when developed or permanent ports and unimproved ports are degraded or inadequate. LOTS operations are conducted by task-organized terminal battalions with Army watercraft units. LOTS operations provide commanders options to extend the mission to improved or unimproved beach sites when anti-access or area denial activities threaten or hinder operations at permanent ports. LOTS also provides a means of intratheater sealift to move forces, equipment, and sustainment cargo closer to tactical assembly areas. Plans to conduct LOTS operations should include site selection and area layout, type of lighterage required, and task organization requirements to operate the LOTS operation. See Figure 6-1 on page 6-2 and Figure 6-2 on page 6-3 for depictions of example LOTS operations.

6-3. Planning considerations for LOTS operations include coordination for engineer support and security, anchorage locations and facilities to offload cargo, MHE or CHE, the marshalling area, and LOCs ingress and egress.

6-4. The first step when planning to open a LOTS site is to determine beach areas available. Primary to this determination is the degree of dispersion attainable, which directly correlates to the daily tonnage requirement and the size of the assigned area. Reconnaissance is required to determine the sites most suitable for operations. A physical as well as a hydrographic reconnaissance should be performed whenever possible to determine proposed beach landing sites. See below for details on site reconnaissance. The selection of site locations should be based on the existing capability to accommodate the desired tonnage. Other major factors that should be considered include:

- Beach gradients.
- Characteristics of the bottom and beach surface.
- Anchorage areas.
- Topographic features.
- Tide and tidal current.
- Surf.

6-5. Weather and oceanography can have a significant impact on LOTS operations and should be considered when developing a plan. Adverse weather conditions can generate beach erosion and create hazardous surf conditions that degrade lighterage capabilities. Hazards to lighterage and discharge facilities increase when wave breakages are high, occur frequently, and have wide surf zones. A breaker period is the time elapsed between each successive wave crest. These hazards can be mitigated at times by use of a causeway floating pier that allows cargo discharge beyond the surf zone, reducing breaker impacts.

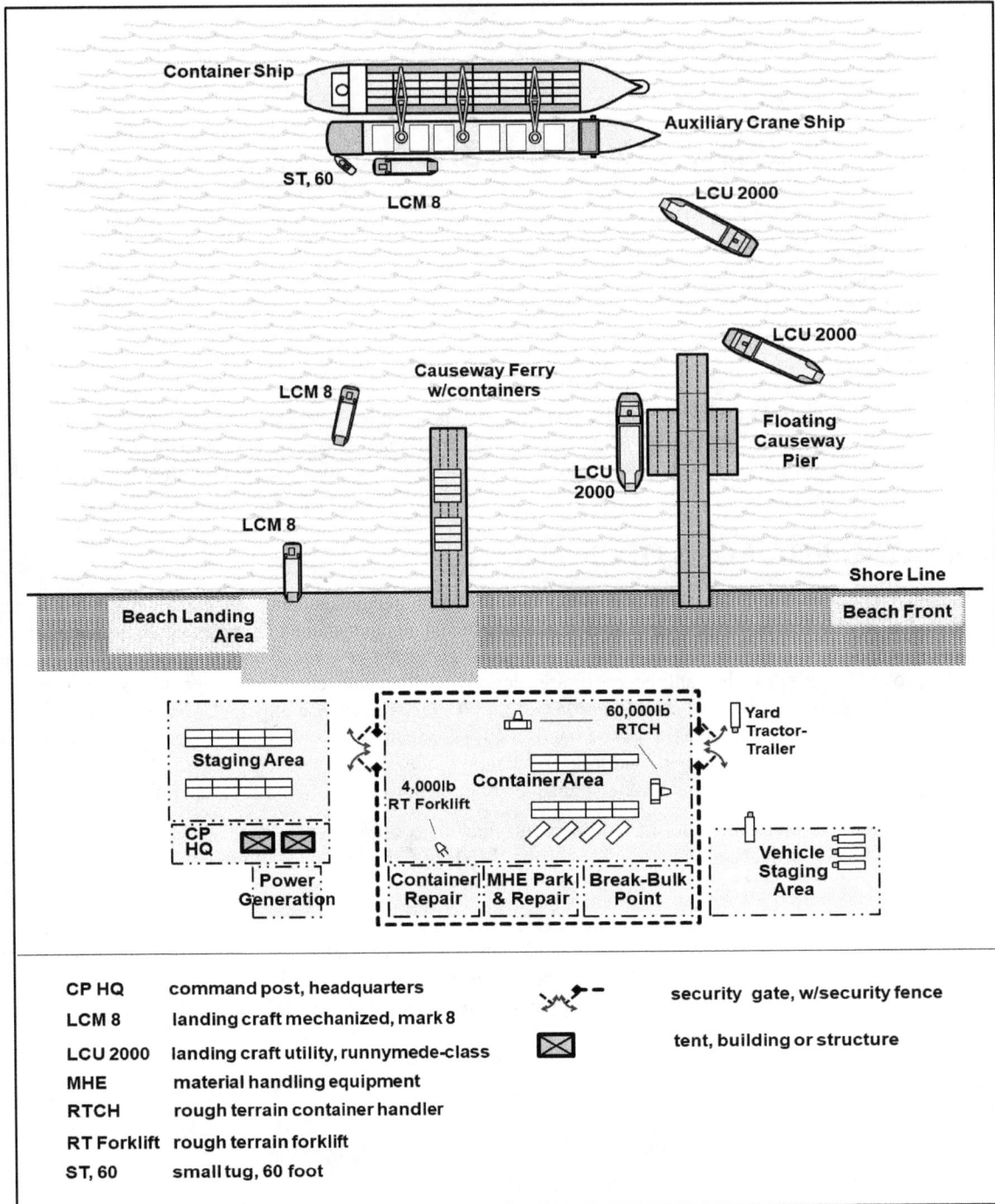

Figure 6-1. LOTS operation example

The figure includes the following labels:

Container Ship

Auxiliary Crane Ship

ST, 60

LCM 8

LCU 2000

LCU 2000

Causeway Ferry w/containers

Floating Causeway Pier

LCM 8

LCU 2000

LCM 8

Shore Line

Beach Landing Area

Beach Front

Staging Area

60,000lb RTCH

Yard Tractor-Trailer

4,000lb RT Forklift

Container Area

CP HQ

Power Generation

Container Repair

MHE Park & Repair

Break-Bulk Point

Vehicle Staging Area

Legend:

CP HQ	command post, headquarters
LCM 8	landing craft mechanized, mark 8
LCU 2000	landing craft utility, runnymede-class
MHE	material handling equipment
RTCH	rough terrain container handler
RT Forklift	rough terrain forklift
ST, 60	small tug, 60 foot

security gate, w/security fence

tent, building or structure

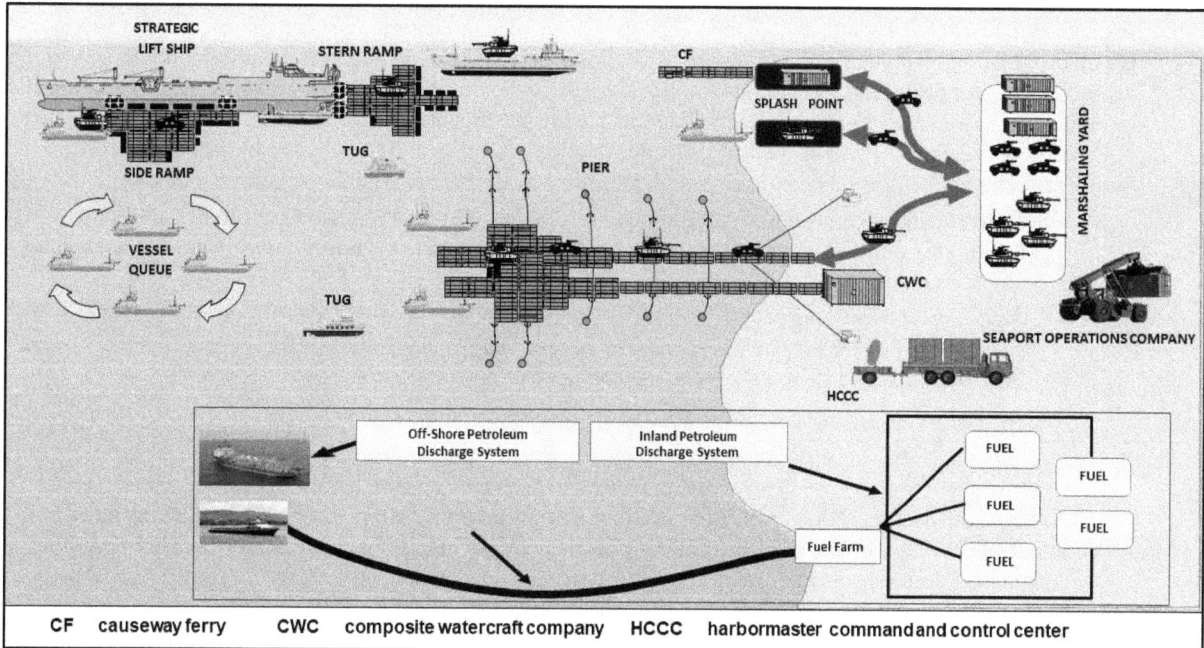

| CF | causeway ferry | CWC | composite watercraft company | HCCC | harbormaster command and control center |

Figure 6-2. LOTS operation example

RECONNAISSANCE AND SITE SELECTION

6-6. The terminal commander normally consults with naval authorities to assess possible beach sites for LOTS operations. This assessment is usually conducted by extensively studying maps and analyzing aerial reconnaissance reports. Detailed ground and water reconnaissance can provide more information on the terrain at the site location and should be conducted whenever feasible. A physical reconnaissance will validate current conditions of road networks, bridge capacities, and any beach or surf limitations. Assigned engineer units will accomplish most major beach preparations.

6-7. A reconnaissance team should contain individuals with the subject matter expertise required to assess the physical characteristics of the location and provide a detailed report to the commander for decision. A reconnaissance team should consist of engineers, military divers, and individuals with expertise in the following areas:

- Terminal operations (preferably a terminal unit representative).
- Meteorology.
- Signal.
- Movement control.
- Intelligence.

6-8. Physical characteristics of the proposed LOTS location considered by the reconnaissance team include—

- Access to LOCs.
- Engineering effort required to prepare and maintain the area.
- Signal construction and maintenance required for communication within the beach area and between the beach area and the terminal headquarters.
- Environmental considerations, including location of beach dumps, transfer points, and maintenance areas.
- Lighterage types that could be employed.
- Anchorage area locations.
- Spud (self-elevating, non-propelled) pier potential and other special equipment requirements.
- Rail network potential.

6-9.   In addition to assessing the beach area itself, the reconnaissance team should also determine if the selected beach area has enough anchorage to accommodate the number and types of ships required to support planned operations. Typically, the numbers of anchorages allowed are based on weather conditions, water depth, underwater obstacles, surf conditions, tidal ranges, and currents. Distances offshore of up to 2 nautical miles is considered acceptable to anchor and conduct efficient discharges. For example, sandbars or reefs just offshore may preclude the use of landing craft, mechanized; landing craft, utility; or barges in certain areas. Significant factors that should be considered to determine available or suitable anchorage include:

- **Depth.** Large cargo ships require a mean low water of 30 feet and a maximum of 210 feet. A fast sealift ship requires a mean low water of 37 feet. The maximum draft of ships to be discharged and the ground swells dictate the minimum depth.
- **Size.** For planning purposes, the anchorage area should be a circle with an 800-foot radius to provide a safe, free-swinging area. This is required for the standard five-hatch vessel. Use the following formula if larger vessels are anticipated in the operation:

$$2(7D + 2L) = \text{diameter in feet.}$$

$$D = \text{depth of water in feet} \qquad L = \text{length of vessel in feet}$$

Note: A much larger radius maybe required for dispersion if operations are being conducted under threat of nuclear warfare or if hazardous materials are included. Bow and stern mooring is not considered desirable in tidal areas because crosscurrents excessively strain mooring gear. An appreciable change in depth also requires continuous watching of the anchored vessels. The type of offshore bottom also significantly affects how close ships can be anchored to each other. A ship will drag anchor if the bottom is too rocky or soft.

- **Landmarks.** Landmarks that assist navigation and location of beaches (such as prominent hills) can be helpful.
- **Underwater obstacles.** Note any underwater obstacles such as bars, shoals, reefs, rocks, wrecks, and enemy installations that might interfere with the passage of vessels to and from the area. Estimate the degree of interference offered and the amount of work involved in clearing channels.

6-10. In selecting a beach for operations, the reconnaissance team should also consider the availability of road or rail networks or the possibility of building one to tie the beach exits to the main transportation network. The reconnaissance team should also consider the requirements of a communications network that may require telephone lines or the ability to reach satellite capabilities. They should evaluate road networks to assess physical characteristics, including the grade of road to determine load capability and strength and the width of any bridges. Finally, they should evaluate the availability of inland waterways. If inland waterways are available, the reconnaissance team considers the potential locks, dams, bridges, or other structures that could impede LOC operations. See JP 3-34, ATP 3-34.5, and FM 3-34 for more information on selecting LOTS locations and environmental considerations.

BEACH CAPACITY

6-11. *Beach capacity* **is the per day estimate expressed in terms of measurement tons, weight tons, or cargo unloaded over a designated strip of shore.** Several factors must be considered to determine the capacity of a particular discharge site. These factors can be divided into the following three groups:
- Those that limit the discharge rate from the vessel in-stream.
- Those that limit the cargo-handling capacity of the beach.
- Those that restrict the flow through the area because of the nature of the beach and the adjacent operational area and infrastructure.

6-12. The group of factors that most limit the quantity of supplies that can be handled determines the capacity of the beach. Beach terminal planning requires making a beach capacity estimate. It involves the same steps that are used in planning for a permanent marine terminal.

FACTORS AFFECTING HANDLING CONSIDERATIONS

6-13. Factors affecting cargo-handling capacity include the following:

- Numbers and experience of personnel available for discharging ships and handling cargo on the beach and in the discharge areas.
- Type and availability of MHE and transportation assets for beach clearance.
- Types and amounts of lighterage available for operations.
- Enemy's ability to interrupt operations.

BEACH TRANSFER POINT

6-14. The requirement for beach transfer points must be considered during the reconnaissance and their locations should be designated. A desirable beach transfer point should include the following:

- Located at the rear of the beach to reduce interfere with shoreline operations.
- Located near the clearance route to allow cargo trucks to receive loads and exit with limited interference of other operations.
- Located near railheads when possible if the node is active.
- Have adequate space between MHE operators loading trucks and amphibian operations.
- Ability to have cranes located on firm, level ground. The crane's longer axis should be parallel to the direction or movement of the vehicles.

TRAFFIC CONTROL

6-15. Traffic control is vital to preventing congestion in the terminal area and promptly clearing cargo to its next destination. An MCT can be employed at this node to provide traffic management of vehicles operating in and around the terminal. Personnel from an ICTC or SOC can also be utilized to document and provide ITV of the cargo. To control vehicle traffic in a beach area, ensure the following:

- A sufficient number of drivers, MHE, and supervisors should be available during all operations.
- A traffic system is established to maximize staging area and minimize congestion of transiting vehicles.
- Communications network to coordinate staging and loading of vehicles.

BEACH EXITS

6-16. The SOC managing the discharge beach operation should limit congestion and move equipment to inland destinations as rapidly as possible. Cluttered beaches increase the possibility of enemy threats and limit throughput capability. The number of exits required varies according to the physical characteristics of the roads, the type and amount of cargo to be handled, and the types of conveyances to be used in beach clearance. Tracked and wheeled vehicles should have separate routes.

6-17. During unloading operations, SOC personnel should be alert for new ways to expedite cargo movement. Each site should have at least one truck dispatcher when clearance is being done by trucks. Two practical expedients are discussed as follows:

- Normally, rough terrain cranes are needed at the shoreline when cargo must be lifted from landing craft and placed in highway transport equipment.
- Floating causeways, RO/RO platforms, and causeway ferries are used to ensure motor vehicles safely reach the beach. They will also eliminate the possibility of drowning out because vehicles can roll ashore without passing through the water.

6-18. The capacity of the road network from the beach to principal inland areas often limits the capacity of a beach to discharge and clear supplies and personnel to inland destinations. The useful capacity of the beach can never exceed the capacity of the road network. Therefore, an early and detailed analysis must be made to determine the capacity of the existing of all potential road networks. If the capacity is inadequate, adjustments may be necessary to reinforce the infrastructure, requiring engineer support for construction and maintenance.

Unit Employment During LOTS

6-19. Within a LOTS operating area, each two-ship terminal is under the direct operational supervision of a terminal battalion. Each terminal is manned by one SOC and lighterage units commensurate with the workload and environment. Motor transportation capability may be required to support cargo clearance. ICTCs may be required to support discharge operations. Harbormaster detachments may also be attached as required. A TBX coordinates the functions of a number of these terminals dispersed along a shoreline. Maintenance for lighters and other Army watercraft is provided by the composite watercraft company. Due to operational requirements, LOTS operations may be dispersed over extended distances. Employing terminal units over widely separated distances along a coastline requires development of a communications, maintenance, and personnel support plan with the capability to support multiple locations.

LOTS OPERATIONS

6-20. LOTS operations are dependent on geographic, tactical, and time considerations that extend from initial operation planning through the deployment of LOTS forces and equipment to ensure the off-load and delivery of cargo. LOTS operations have the ability to create pier facilities, conduct salvage, or provide floating crane support alongside ships or permanent facilities. The types of LOTS operations include bare beach or degraded port, unimproved facility, or augmentation of a permanent port.

Bare Beach or Degraded Port

6-21. During bare beach or degraded port operations, lighterage is used to off-load ships in stream (at anchor) and cargo is moved over a degraded beach or onto the shore. Bare beach operations are utilized when port facilities are not available, damaged, have been denied, or cannot accommodate adequate throughput. LOTS operations require secure routes that are established to and from the beach. Engineers may be required to ensure movement can occur through the surf zone, across the beach, and into marshaling yards or onto clearance transportation.

Unimproved Facility or Augmenting a Permanent Port

6-22. An unimproved water terminal site is not specifically designed for cargo discharge. The site lacks the facilities, equipment, or infrastructure of a permanent water terminal. An unimproved facility may lack sufficient water depth, pier length, and MHE. Additionally, the port may have sustained damage, making it equivalent to an unimproved facility. LOTS can provide augmentation when an undamaged or otherwise functional port is insufficient to support the capability or throughput required.

Shore-to-Shore Operations

6-23. Tactical and logistical shore-to-shore operations may be conducted across or along rivers, between islands, along a coastline, or between a continental land mass and an offshore island. Except for the fact that ocean shipping is not involved, terminal unit functions in these operations are the same as described for bare beach operations. The units provide the same support as described in previous chapters. Command elements and relationships in logistical shore-to-shore operations are the same as in conventional marine terminals and in ship-to-shore bare beach LOTS operations. The SOC ship platoons work in the loading area on the near shore, and the shore platoons operate discharge points in the objective area. Landing craft units provide the lighterage service. Terminal elements may be assigned to clear cargo backlogs.

Chapter 7

Transportation Brigade Expeditionary

This chapter discusses the TBX, its role, and its ability to provide terminal operations support.

OVERVIEW

7-1. The TBX is designed to manage terminal operations performed by military, contracted, or HN labor forces and provides cargo documentation services and ITV of equipment and cargo freight transiting seaports. The TBX provides global command and control of Army watercraft and water terminal capabilities and organizations. The TBX is under the operational control of Army Forces Command and is normally attached to a TSC or ESC. The TBX serves as the TSC or ESC commander's primary expert on port operations and management.

CONCEPT OF OPERATIONS

7-2. The TBX is a brigade-level headquarters capable of providing mission command of assigned and attached water terminal and watercraft units engaged in conducting deployment, redeployment, and distribution support (see Figure 7-1).

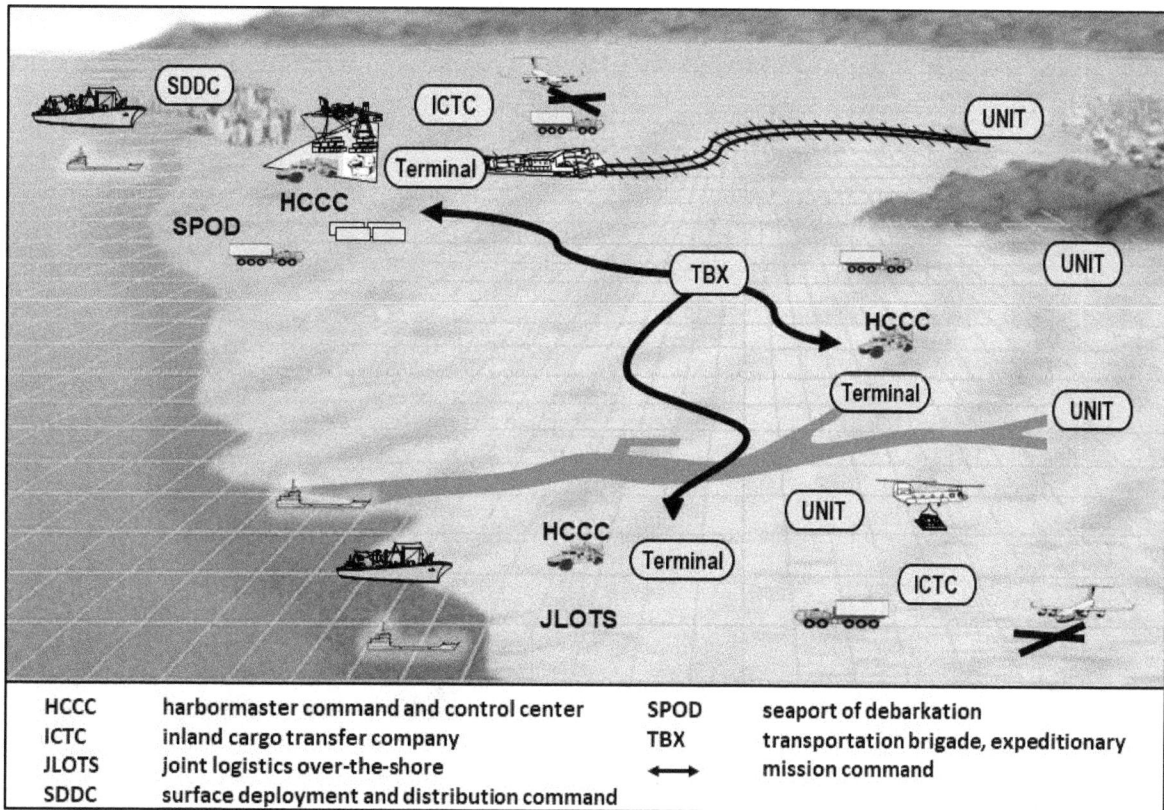

HCCC	harbormaster command and control center	SPOD	seaport of debarkation
ICTC	inland cargo transfer company	TBX	transportation brigade, expeditionary
JLOTS	joint logistics over-the-shore	⟷	mission command
SDDC	surface deployment and distribution command		

Figure 7-1. Transportation brigade expeditionary employment

MISSION COMMAND

7-3.   The TBX provides mission command of assigned and attached port, terminal, and watercraft units conducting expeditionary intermodal operations in support of multidomain operations. The TBX will normally be placed under the operational control of an ASCC and attached to the appropriate headquarters for mission command when deployed. Water terminal and watercraft units assigned to the TBX conduct deployment, redeployment, and distribution support IAW CCMD (command authority) and operational requirements. The TBX and its subordinate battalions establish and maintain close coordination with the TSC or ESC and the sustainment brigades responsible for executing the theater distribution mission. When attached to a TSC or ESC, the TBX will establish the same close mission coordination with the SDDC single port manager and port commanders. The TBX's ability to maintain close mission coordination between SDDC and the TSC or ESC will ensure a seamless theater strategic-to-tactical transition from port opening to distribution operations in a manner that meets CCDR operational priorities.

7-4.   During the build-up of combat forces in a theater of operations, the TBX rapidly deploys sufficient command and staff capability to provide mission command for port opening operations (see Figure 7-2). The organizational structure required to execute the theater opening function is dependent upon mission variables, and the size and makeup of the TBX command team must be tailored to meet the operational requirement during early deployment operations. The TBX rapid port opening command team must be capable of establishing the port operating site; holding, staging, and marshaling areas; life support; and distribution operations. The command team must also be sufficient to control multiple battalions engaged in port opening, water terminal, and watercraft operations. TBX essential mission tasks include—

- Deploy to a theater of operations.
- Establish the TBX operational area.
- Plan and manage watercraft and water terminal support for theater opening.
- Conduct port opening operations.
- Conduct water terminal and watercraft operations.

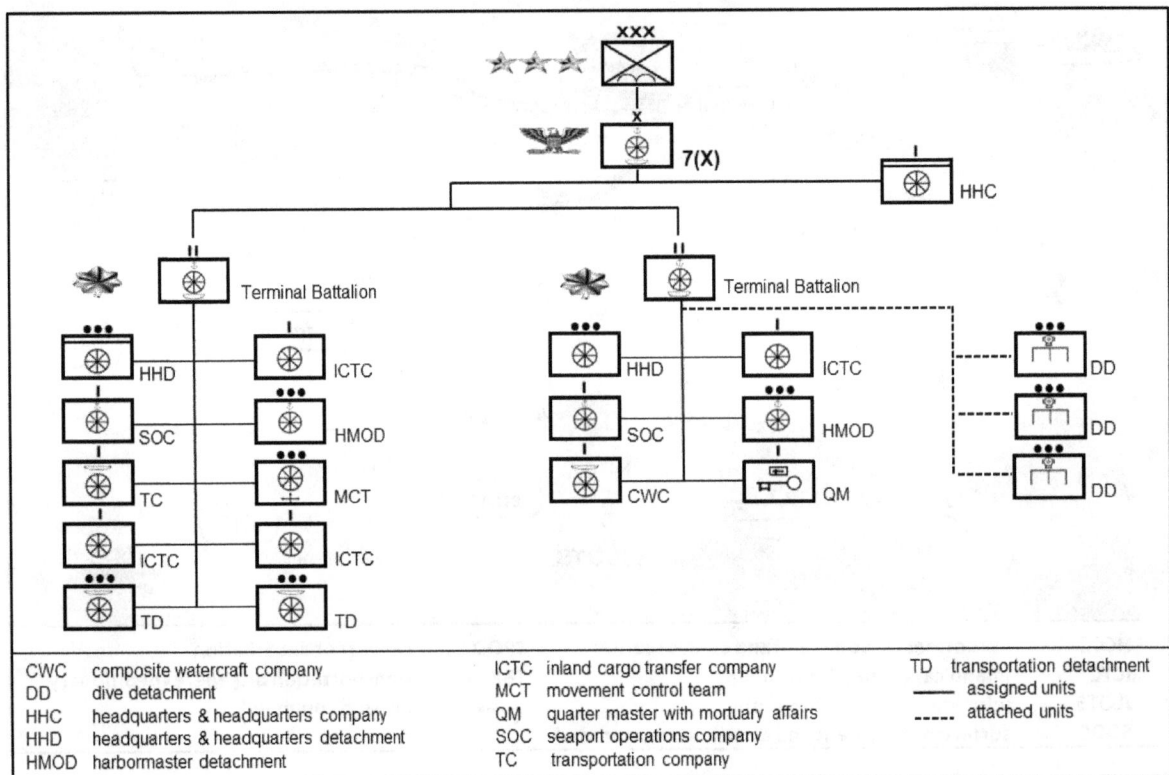

Figure 7-2. Transportation brigade expeditionary task organization

REQUIRED CAPABILITIES

7-5. The TBX provides mission command of units engaged in water terminal and waterborne distribution operations, and planning and management of water terminal and watercraft capabilities. The TBX staff is capable of providing—

- Scalable, rapidly deployable command team capable of providing command and control for rapid port opening operations.
- Mission command of up to seven terminal battalions when fully deployed.

7-6. The TBX is capable of providing technical supervision of up to seven terminal battalions when fully deployed. The TBX staff is specifically trained to conduct port opening operations to receive, load, discharge, stage, maintain control and ITV of, and release equipment and materiel to the receiving unit or command. The TBX is capable of deploying to and operating in all SPOEs/SPODs. Ideally the SPOE/SPOD is a well-equipped, permanent facility capable of discharging large, medium-speed RO/RO ships. However, the port can be a permanent facility capable of discharging a variety of vessels, an austere port requiring ships to be equipped with the capability to conduct their own offloading, or beaches requiring LOTS operations.

7-7. TBX critical wartime and implied missions include—

- Plan and manage watercraft and water terminal support for a theater of operations.
- Conduct water terminal and watercraft operations.
- Implement and monitor theater port operations.
- Commit terminal and watercraft assets in support of theater deployment and movement operations.
- Monitor and maintain status of terminal and watercraft assets to ensure they are properly employed and not over-tasked.
- Provide terminal infrastructure assessment.
- Monitor and coordinate operations and positioning of water terminal and watercraft assets in theater.
- Provide operational control, administration, logistics, and supervision of assigned and attached units.
- Assist in the coordinated defense of the unit's area or installation.
- Plan and conduct redeployment operations.
- Redeploy to another contingency or home station.

7-8. The TBX provides theater opening capabilities through mission command of assigned and attached port, terminal, watercraft, mortuary affairs, and movement control units conducting expeditionary intermodal operations in support of multidomain operations.

7-9. The TBX deploys to a theater of operations to provide mission command for port opening and operation of inland waterway, bare beach, degraded, and improved sea terminals in support of CCMD (command authority) theater opening operations. The headquarters is organized to provide the ability to rapidly deploy minimum capabilities to meet requirements for rapid port opening operations and small-scale contingencies, while maintaining a stay-back structure to conduct ongoing peacetime planning and perform mission command functions.

7-10. In the event of sustained large-scale operations, the TBX can be fully deployed to provide mission command for Army water terminal and watercraft operations in any operational environment. Regardless of the operational environment or scale of the operation, TBX primary tasks are to rapidly deploy; establish and maintain port operations; establish and coordinate terminal protection operations; conduct waterborne distribution and LOTS operations; conduct joint reception, staging, and onward movement of cargo; establish and coordinate life support services and contract management for terminal operations; and provide container management and joint documentation oversight. The TBX is capable of providing mission command of units organized under tables of organization and equipment and TDAs (see Figure 7-3 on page 7-4).

Expeditionary Capability

Provide **Theater Opening** capabilities through **mission command** of assigned and attached **port, terminal, watercraft, movement control,** and **mortuary affairs** units conducting <u>**expeditionary intermodal operations**</u> in support of **multidomain operations**.

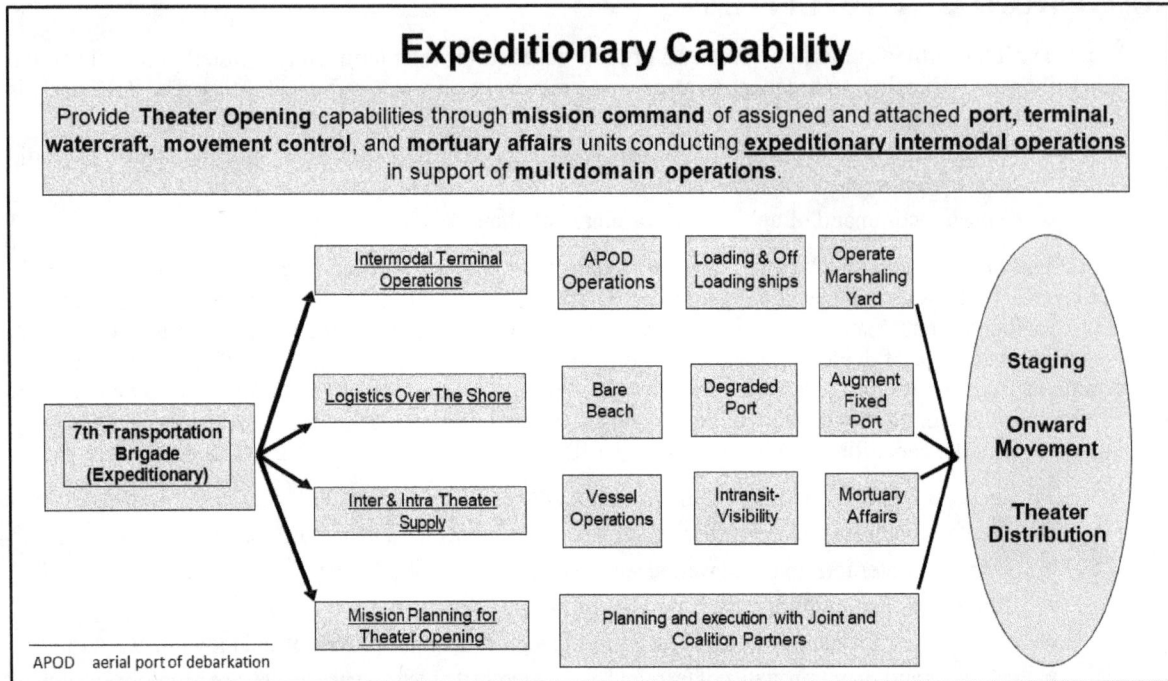

Figure 7-3. Transportation brigade expeditionary capabilities

GLOBAL REACH

7-11. The TBX staff maintains oversight of the Army's globally dispersed water terminal and watercraft organizations and capabilities, including those not directly under the TBX command. The TBX must develop relationships and communications with the various commands to which Army water terminal and watercraft units and assets are assigned or attached in order to maintain visibility of current status and operational readiness. In this role, the TBX becomes the repository of subject matter expertise on the current status, capabilities, and proper employment of Army water terminal and watercraft capabilities.

7-12. The TBX staff will establish and maintain relationships and communications with CCDR operational planners, exercise executive agents, global force providers, and Army prepositioned fleet managers in order to understand and to provide operational advice in planning for and employing these assets and capabilities in exercises and contingency operations. The key functions performed by the TBX in this role include—

- Knowing the capabilities and current status of water terminal and watercraft units in the Army.
- Knowing the capabilities of water port complexes worldwide (infrastructure and civilian workforce capabilities).
- Seeing, early on, requirements for these capabilities that are identified by CCMDs.
- Assisting CCMDs in determining the true requirement in the technical area of maritime terminal operations and watercraft employment.
- Quickly matching requirements to available capabilities.
- Making timely recommendations for force allocation, including use of non-Army capabilities.
- When given the authority, allocating Army forces or procuring non-Army water terminal and watercraft capabilities.

7-13. Army watercraft systems under the mission command of the TBX enable operational agility through the delivery and sustainment of operationally significant forces to the point of employment in support of multidomain operations. These capabilities deliver—

- Combat enablers that bridge theater strategic deployment with operational and tactical employment of ground combat forces.

- Organic Army capabilities designed for and focused on moving, maneuvering, and supporting sustained land operations.
- Land forces with operational reach and agility through tactically synchronized movement of combat-ready, tailored formations dispersed across the depth of the operational environment.
- Increased tactical agility by exploiting the littoral boundaries to expand access, mitigate anti-access/area denial environments, and create multiple dilemmas for the adversary.

7-14. Unless fully deployed, the TBX staff must provide a stay-behind command team to maintain command and control and technical supervision of assigned and attached units that are not deployed. The organizational structure required to execute the stay behind mission will be partially dependent on the staffing requirements for the deployed command team. However, the TBX commander must ensure sufficient staff expertise is retained to maintain command and control of non-deployed units. The TBX stay-behind command team will also provide reach-back capability for the deployed command team, enabling staff rotations and planning support for changing operational requirements in theater. Other responsibilities include—

- Develop interoperability with joint and multinational partners to exercise and enhance combined joint LOTS capabilities.
- Leverage links and mutual capabilities with the Army Reserve at the forefront of total force integration.
- Execute a vital role supporting mobility and maneuver, access options, deployment of forces, and sustainment through austere access points.

This page intentionally left blank.

Appendix A
Marshaling Yard Operations

The purpose of this appendix is to discuss marshaling areas and their role in intermodal operations. Use of a marshaling area allows rapid clearing of the beach or pier. It makes vessel working space available and reduces pier congestion, thus reducing the potential for work slowdowns or stoppages in discharge operations.

CARGO OPERATIONS

A-1. Ideally, containers and other cargo are transferred directly onto line-haul equipment for movement inland. In most cases, this is not possible except for selected containers or other cargo. Conceptually, all cargo should move through the terminal without delay. However, this is not always possible because of the following:

- The consignee's reception capacity may be limited.
- The movement plan causes delays in clearance.
- Damaged containers may require repair or stowing of contents before further movement.
- Containers may require segregation by destination or priority.
- Containers occasionally require re-documentation before further movement.
- Some retrograde containers must be cleaned and fumigated.
- Containers found with broken seals or apparent pilferage must be inventoried and new seals applied before onward movement.

A-2. The marshaling yard is a temporary, in-transit storage area. It expedites discharge operations by facilitating rapid and continuous movement of cargo and containers to or from the beach or pier. Marshaling cargo levels the line-haul peak workloads that result from discharge operations. Concurrently, marshaling cargo allows selective, controlled, and flexible phasing of container or cargo movement to destination or vessel.

ORGANIZATIONS AND FUNCTIONS

A-3. A marshaling yard has no set organization or physical layout. It is organized to meet operational requirements within available space. By grouping related functions, the design of the marshaling yard will eliminate lost motion, reduce container and cargo handling requirements, and permit a logical flow of containers and cargo through the terminal. Cargo can be subdivided into any number of categories. The most widely used categories are general (breakbulk), containerized, RO/RO (vehicles and containers on chassis), and special (oversize, heavy lift, hazardous, and security). These categories and the volume of cargo in each category play a significant role in marshaling yard organization. All terminals should provide for the following activities and functions:

- A central control and inspection point with multiple lanes for cargo and containers entering or leaving the marshaling yard.
- Auxiliary internal checkpoints for containers and cargo entering the yard from a beach, rail spur, or heliport within the yard.
- A traffic circulation plan depicting movement flow into, through, and out of the marshaling area.
- Segregation of inbound containers and cargo by size and type. Within these groupings, further segregation by destination, priority, and special handling requirements (security, mail, and hazardous materials).
- Segregation of retrograde cargo and containers by type and size with empty and loaded containers further segregated.

● Running inventory of containers by location and status within the yard.

● Security area for breakbulk or containerized sensitive and high-dollar-value cargo.

● External power source for refrigerated containers. (Self-contained refrigeration units may be required in an unimproved or bare-beach LOTS environment. This mandates separate propane or diesel refueling areas.) Refrigeration maintenance must also be provided.

● Sheltered facilities for inventory and control, documentation, and movement control elements.

● Covered facilities for stowing containers and repairing cargo.

● Cleaning or decontamination of retrograde containers and vehicles.

● Minor repair of damaged containers.

● Equipment parking.

● Unit maintenance of equipment.

● Messing and comfort facilities.

● Spill contingency plans including emergency supplies and equipment for containing and disposing of hazardous material spills. See TM 3-34.56.

● Disposal of hazardous and special waste IAW federal, state, local, and HN environmental regulations. See TM 3-34.56 and DODM 4715.05, Vol. 1.

A-4. Yard surfacing requirements of existing ports and those under construction are intended to support commercially operated equipment. The load-bearing capacities will need to meet foreseeable requirements. Figure A-1 depicts an example layout for a container marshaling yard in an unimproved or bare-beach LOTS environment.

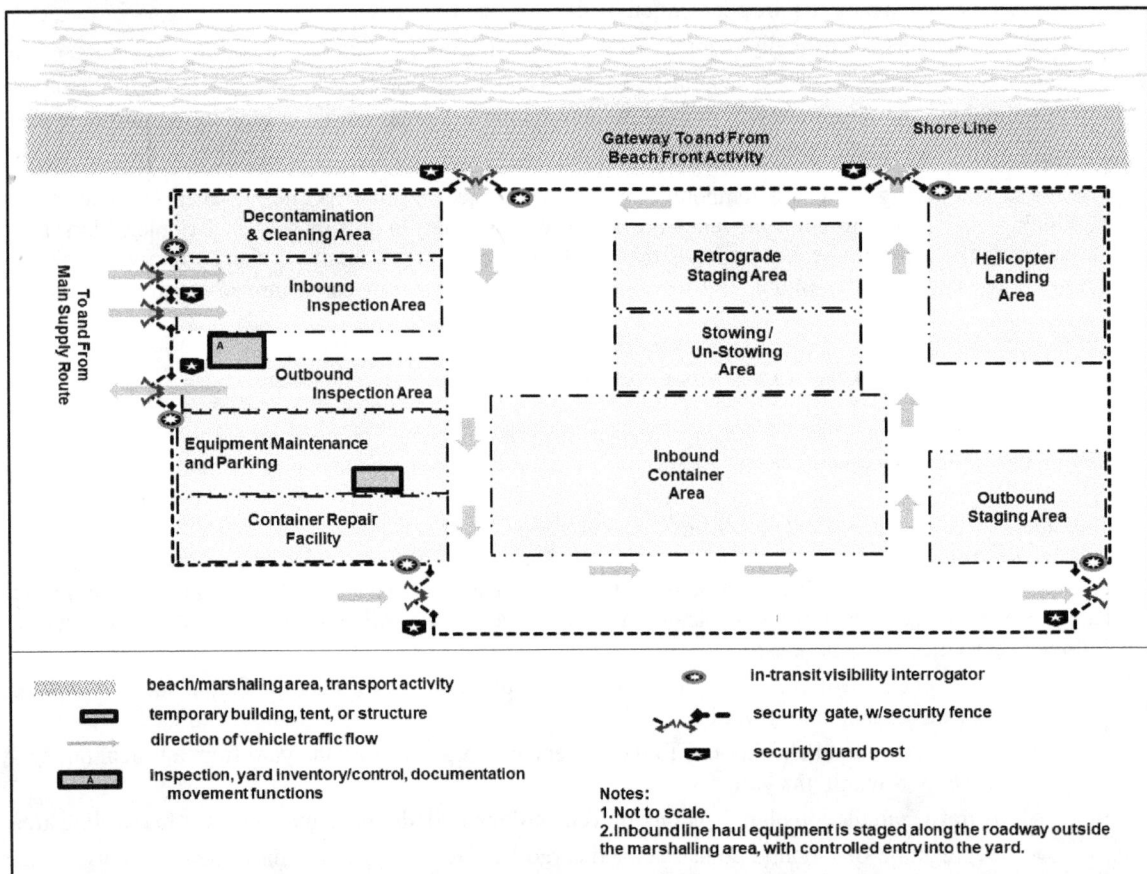

Figure A-1. Example organization for a container marshaling yard in a LOTS environment

LOGISTICS OVER-THE-SHORE OPERATIONS

A-5. In a LOTS environment, the marshaling yard surface may be subjected to loads of about 218,000 pounds (50,000-pound front loader with a 40-foot container). Normally, beach movement of containers would be restricted to 20-foot containers or less. However, containers up to 40 feet may be used. An unimproved or bare beach facility does not normally have any surfaced area. Such surfacing must be provided comparable to that in a permanent or semi-permanent facility. A minimum surface would consist of 9 inches of rock or shell sub grade covered with an equal thickness of blacktop. Time constraints would prevent this type of construction in a LOTS environment. The materials that follow may prove useful to support limited loads in LOTS operations:

- **Matting, AM-2.** This is a Navy-developed, extruded aluminum airfield mat. It is designed to support jet aircraft over soft, fine-grained soil. Because of limited stocks, high cost, and high priority for airfield use, this material will probably not be available for marshaling area use.
- **MOBI-MAT.** This fiberglass-reinforced plastic is rolled out in sections that may be bolted together or overlapped. It is less susceptible to water penetration and more easily placed than metal matting. It is effective over beach sand, granular soils, and some fine-grained soils (clay and silt). It relies on the support provided by underlying soils.

LOCATION OF MARSHALING YARD

A-6. Location of the marshaling yard is key to terminal operations and cargo throughput. The objectives of loading and discharging focus on rapid, efficient, and controlled movement of cargo between ship and shore. Therefore, the closer a marshaling yard is to the water terminal the greater the chance to maximize throughput.

CONTAINER MARSHALING AREA

A-7. The marshaling area (general cargo or container) is located as near the vessel, rail, air, or truck discharge or load site as practicable. Enemy capabilities and activities may require dispersion of activities or may otherwise affect the selection of the marshaling yard location.

PERMANENT AND SEMI-PERMANENT PORTS

A-8. The marshaling yard in an existing port is normally next to the pier area with a sufficient pier apron (100 to 500 feet) between the yard and shipside. These distances accommodate container discharge and container clearance activities and are more than adequate for general cargo operations. Rail spurs, warehouses, and similar facilities usually exist but may require rehabilitation. The semi-permanent port is constructed to replace an unimproved or bare beach LOTS site when a suitable permanent port is not available. Layout and construction of the semi-permanent port parallels that of the permanent port. Construction of the marshaling yard should encompass any existing hardstand structures and rail lines.

LOGISTICS OVER-THE-SHORE (UNIMPROVED FACILITY OR BARE BEACH OPERATIONS)

A-9. The LOTS marshaling yard should be approximately 1/4 to 1/2 mile (.4 to .8 kilometers) inland from the beach or dune area to allow an acceptable rate of beach clearance. The maximum distance should not exceed that needed for operations. LOTS operations are inherently inefficient. They should be used only until permanent facilities can be placed into operation or until semi-permanent facilities can be constructed. Port operational considerations and construction details dictate the length of time LOTS operations continue. Engineers assist in assessing the following factors that influence marshaling yard site selection in a LOTS environment (unimproved facility or bare beach):

- **Accessibility.** Is the area readily accessible from the MSR and from the beach? Are internal road networks adequate? If helicopter operations are anticipated, are there any flight obstructions? Is the proposed site next to existing rail facilities?
- **Physical facilities.** Are usable physical facilities available? Are they served by more than one entrance and exit? Are usable hardstands, airfields, railways or rail spurs, buildings, storage sheds, or warehouses in the area?

- **Adequacy of space.** Will available space hold the type, size, and quantity of cargo and containers programmed for the area? Is there adequate area for working and intersecting aisles? Will available space accommodate administrative activities; repair, maintenance, and decontamination operations; retrograde staging; and storage of handling equipment? Is there sufficient area to stage line-haul equipment pending entry into the marshaling yard for loading?
- **Gradient, drainage, and soil characteristics.** Is the marshaling area sufficiently level, with minor grading, to permit general cargo stacking and two-high container stacking without toppling? Are surface and subsoil drainage adequate? What is the depth and type of subsoil? Is the surface soil compatible? Does the soil need compaction, stabilization, or surface matting?

CONTAINER OPERATIONS

A-10. The configuration of the marshaling yard is determined by container stacking operations. More specifically, the configuration of the marshaling yard considers key factors such as—

- Space available.
- Type of surface (bare beach, asphalt).
- MHE.
- CHE.
- Desired throughput.

CONTAINER STACKING CONFIGURATIONS

A-11. Containers may be placed in the marshaling yard either on chassis or stacked off chassis. Keeping containers on chassis reduces container handling and accelerates operations. However, when containers stay on chassis throughout the system, one chassis for every two to three containers is needed to support the system. Storing containers on chassis also increases space requirements in the marshaling area.

A-12. The Army operational concept is to stack-load containers off chassis, with a maximum of two high, using the turret stacking method. Retrograde empty containers can be stacked five high if this height is within the capability of the CHE. Other space considerations include stacking collapsed flat racks. Flat racks should be stacked as high as possible by available CHE in an area that facilitates retrograde for eventual back-loading. Although stacking containers increases handling, it requires fewer chassis and reduces requirements for marshaling yard space. The primary configurations of off-chassis stacking are ribbon stacking, block stacking, and turret stacking.

Ribbon Stacking

A-13. Ribbon stacking is best utilized when selective extraction of containers from the stack is not needed. This method requires more space than block stacking but is more space efficient than turret stacking. Use the ribbon stacking method if selective extraction is not required. Ribbon stacking may be used when all containers in the stack must be reached from the working aisle (the aisle between ribbons) but extraction of a particular container in the stack is not required. As illustrated in Figure A-2, extraction of container A requires that container B first be removed and placed in the working aisle or carried completely out of the block. This results in increased handling requirements and traffic congestion.

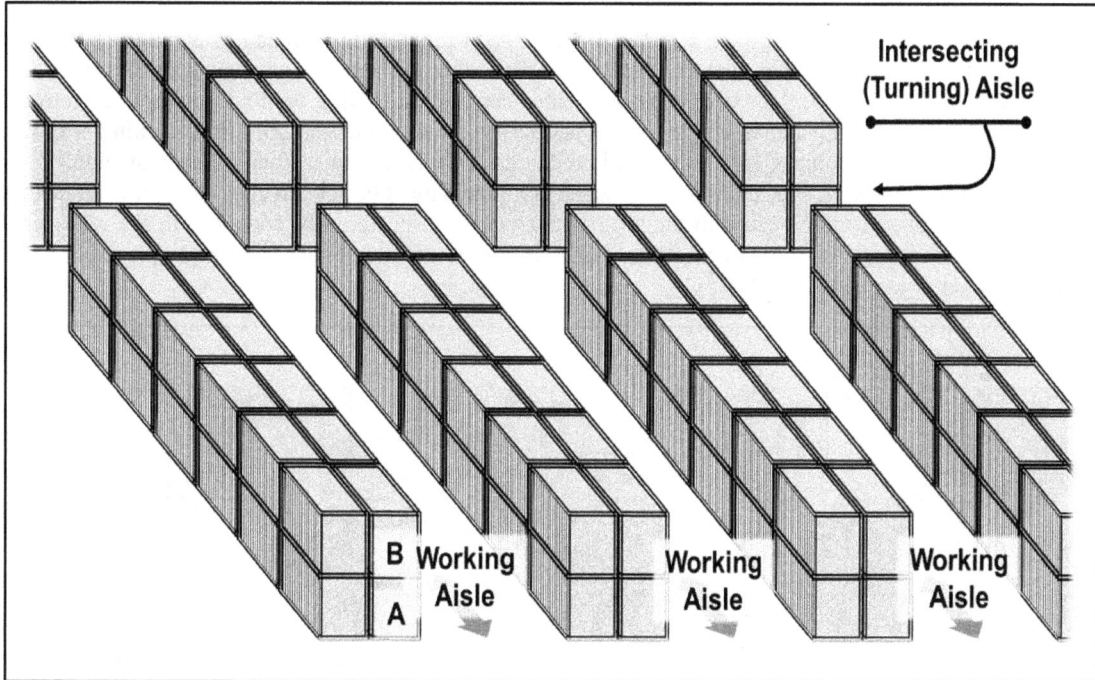

Figure A-2. Container ribbon stacking configuration

Block Stacking

A-14. Container block stacking (Figure A-3) should be used when the containers have a common destination or when selective extraction of containers from the stack is not needed. This method is particularly suited to stacking both empty and loaded identical retrograde containers. It is the most effective use of marshaling yard space. Block stacking is ideal for identical retrograde containers, containers with a common destination, and in other cases where selective extraction is not required. Of the three stacking methods, block stacking uses space most economically.

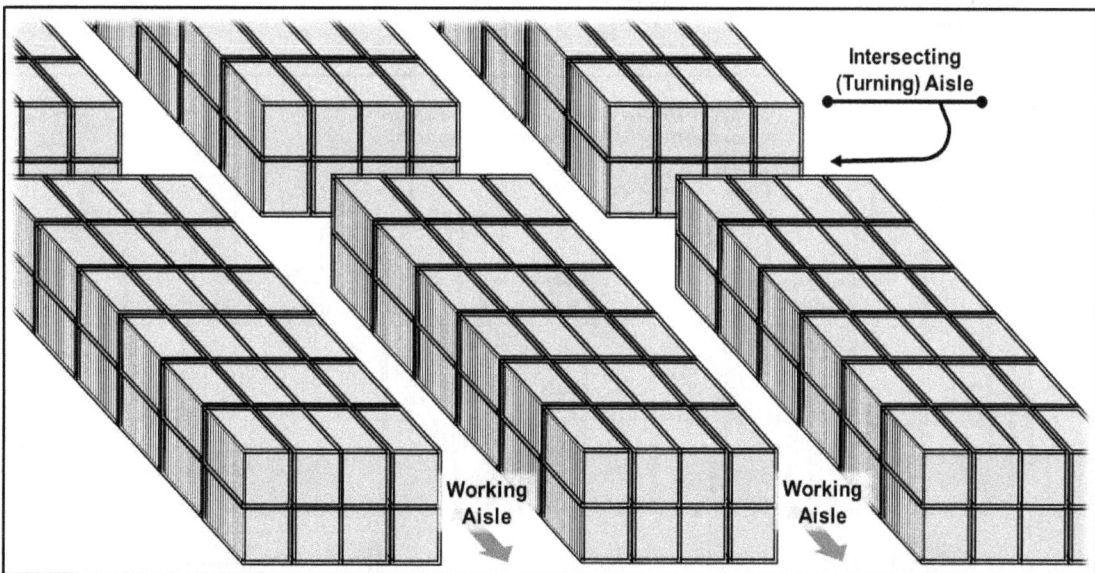

Figure A-3. Container block stacking

Turret Stacking

A-15. Container turret stacking (see Figures A-4 and A-5) requires less container handling for selective container extraction than ribbon or block stacking. Of the three off-chassis configurations, turret stacking least effectively uses space. However, it greatly enhances the marshaling yard's throughput or retrograde operations where selective container-handling is necessary. Although three-high turret stacking is shown in Figure A-5, the Army concept is to stack-load containers only two high. Although turret stacking may be least economical in space, it is recommended when containers must be selectively extracted from the stack. As illustrated, one in three spaces in the second tier remains vacant. Any container in the stack can be removed with no more than two movements. For example, to get to container C, simply place container B over container A to expose container C.

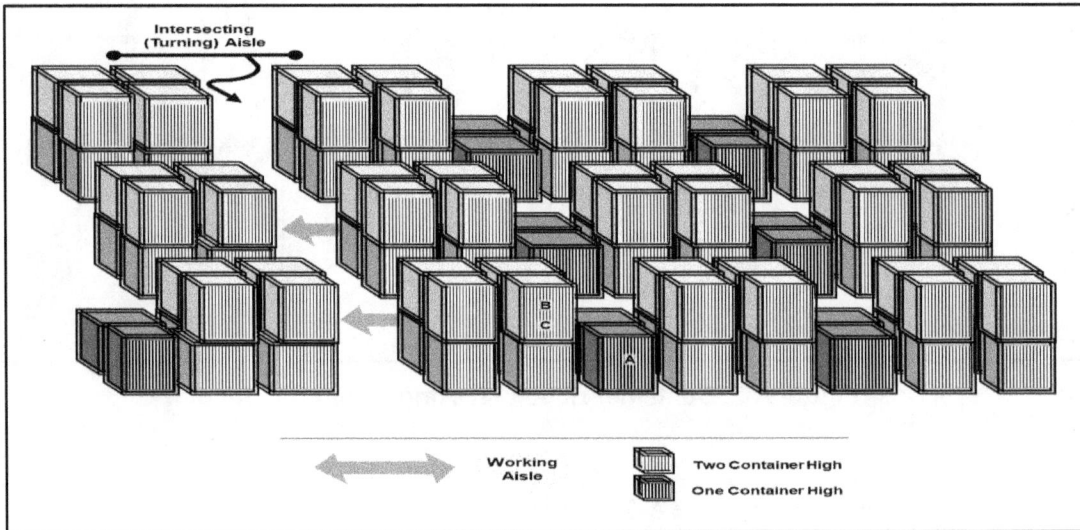

Figure A-4. Container turret stacking (two-high)

Note: In Figure A-5, two spaces out of four remain vacant in the third tier. Any container can be extracted in three or less movements. The method can be used in Army operations only when empty identical containers are being handled.

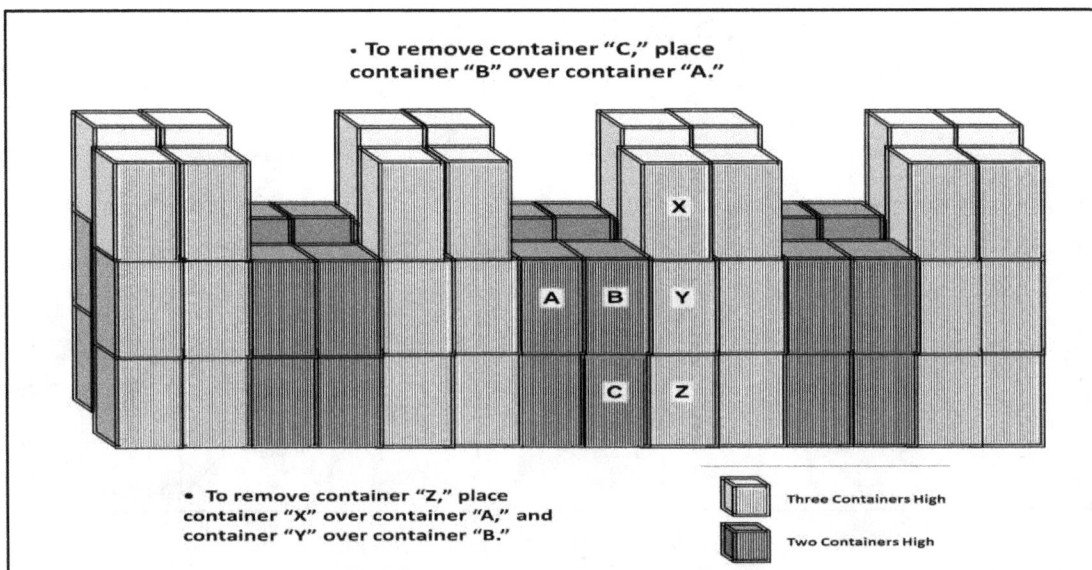

Figure A-5. Container turret stacking (three-high)

A-16. The container-on-chassis marshaling system (see Figure A-6) is most often used in commercial operations. Container-on-chassis marshaling is normally used in marine terminal operations where the container is lifted off the containership directly onto land transport, or in RO/RO operations where the container-on-chassis rig is towed ashore from the RO/RO ship. Marshaling containers on chassis reduces container handling and increases mobility and flexibility of operations. This method increases marshaling yard space requirements. It dictates a 2–1 or 3–1 (or better) container-to-chassis ratio.

Note: Figure A-6 depicts two patterns of container-on-chassis (also referred to as container-on-wheels) marshaling: herringbone (pattern A) and straight-in (pattern B). A commercial operator can significantly reduce container damage in the marshaling area by changing pattern A parking to pattern B. The same operator using pattern B can also, in the case of blacktop surfacing, install a narrow, hardened strip to support the legs of parked chassis. This keeps the legs from sinking into the blacktop.

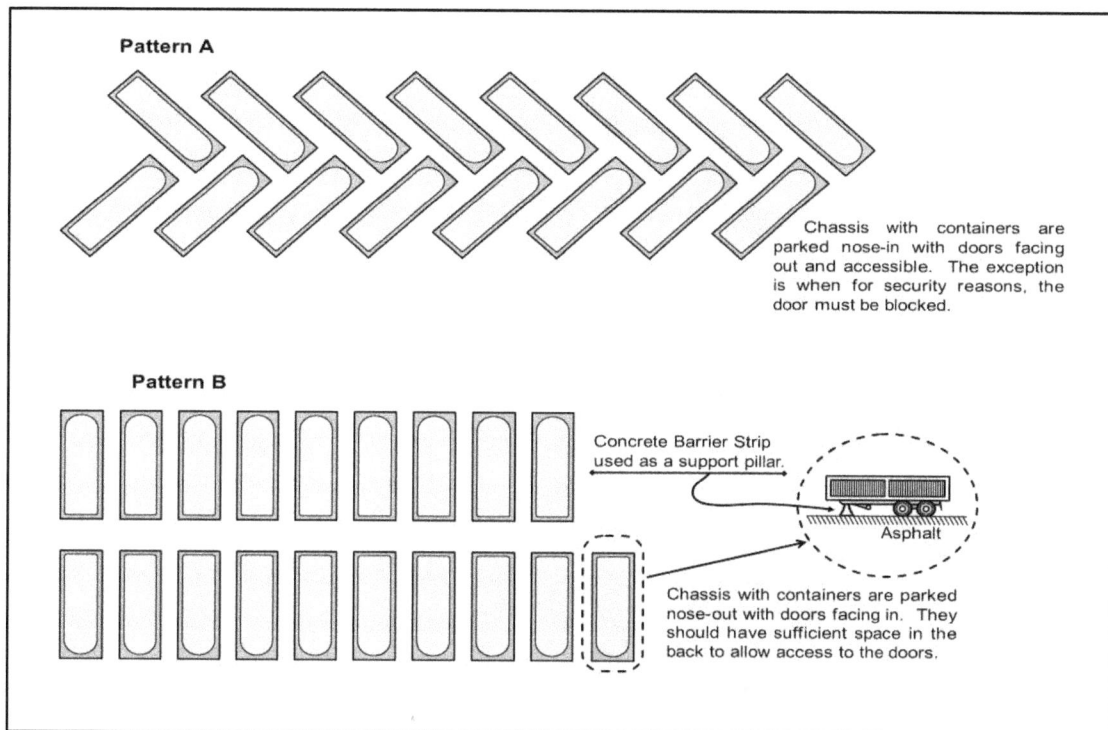

Figure A-6. Container-on-chassis marshaling system

Space Requirements

A-17. Numerous factors and combinations of factors dictate container stacking space requirements. Primary factors include stacking configurations, skill of operator, physical characteristics of CHE, container size, and container cluster concept. Ribbon stacking requires more space than block stacking; turret stacking requires more than ribbon stacking. One-high stacking requires about twice the space of two-high stacking for the same number of containers. The relative space requirements of on-chassis versus off-chassis stacking are obvious.

PHYSICAL CHARACTERISTICS OF CONTAINER HANDLING EQUIPMENT AND CONTAINER SIZE

A-18. The recommended minimum operations space is a 15-foot working aisle with a 50-foot intersecting (turning) aisle when using side loaders. When using a front loader, the overall length of the container being carried determines the effective width of the front loader. For example, with a 20-foot container, the width of the vehicle is 20 feet. In a 90-degree stacking operation, a typical front loader carrying a 20-foot container has a 45-foot turning radius. Aisle width must be adjusted to accommodate different container lengths.

CONTAINER CLUSTER CONCEPT

A-19. Figures A-7 through A-11 present conceptual procedures for computing space requirements to stack containers in a marshaling yard. The concept envisions making clusters of containers grouped as needed to accommodate specific operational requirements or environments. Clusters are developed for turret stacking, block stacking, and on-chassis parking for 20-foot and 40-foot containers. Variations accommodate turret front-loader or side-loader stacking. The intersecting aisles are omitted in figures A-7 through A-10.

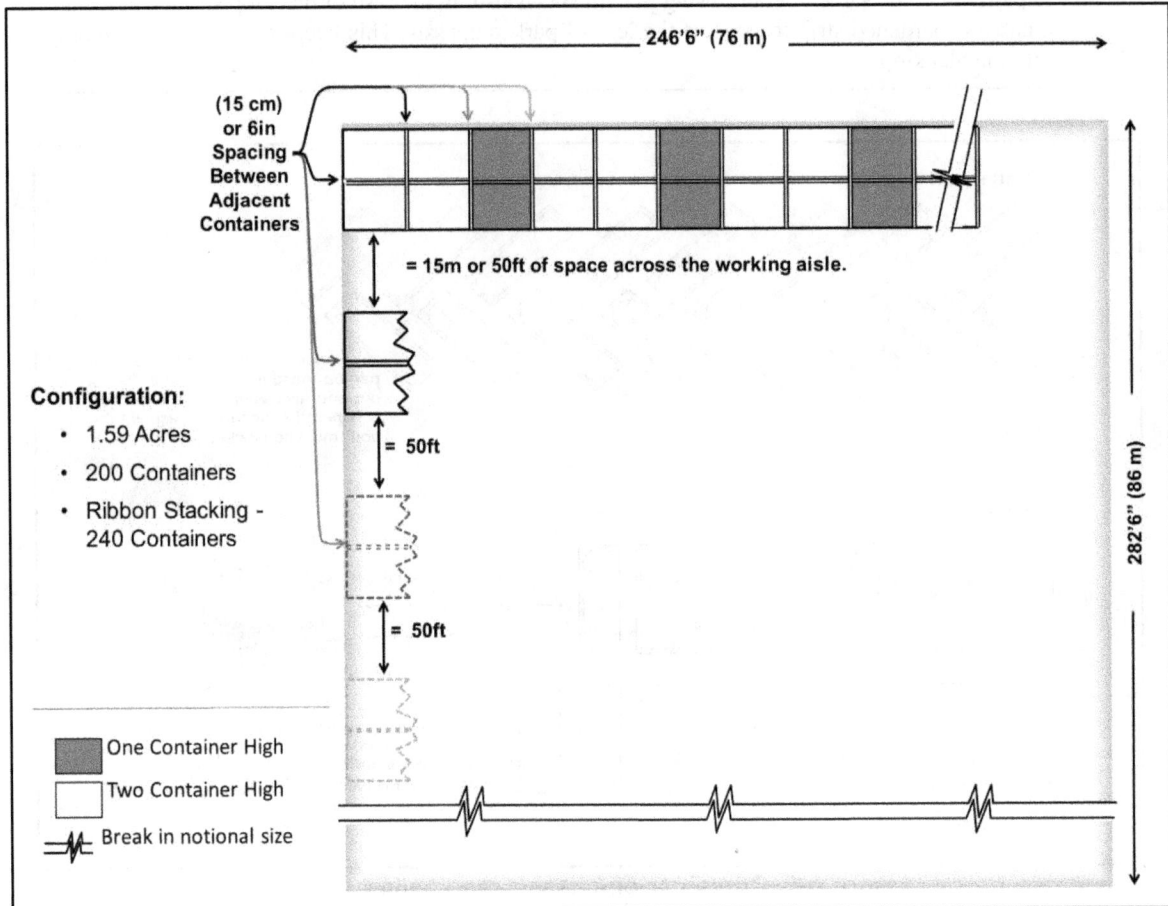

Figure A-7. Cluster plan for front loader turret stacking of 20-foot containers (50-foot working aisles)

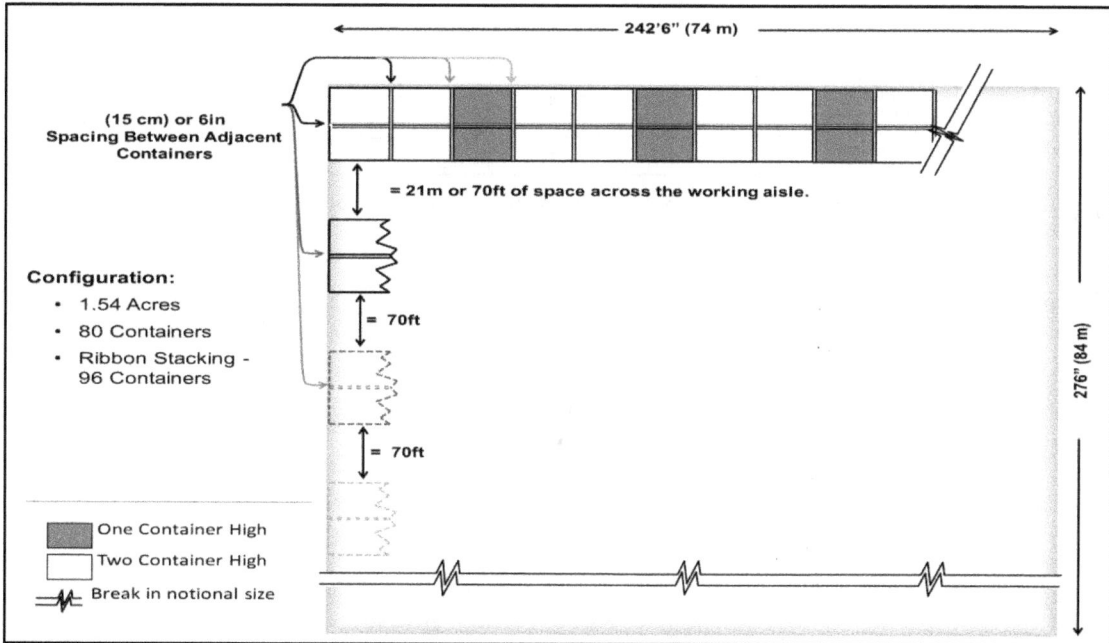

Figure A-8. Cluster plan for front loader turret stacking of 40-foot containers (70-foot working aisles)

Configuration:
- 1.54 Acres
- 80 Containers
- Ribbon Stacking - 96 Containers

242'6" (74 m)

(15 cm) or 6in Spacing Between Adjacent Containers

= 21m or 70ft of space across the working aisle.

= 70ft

= 70ft

276" (84 m)

- One Container High
- Two Container High
- Break in notional size

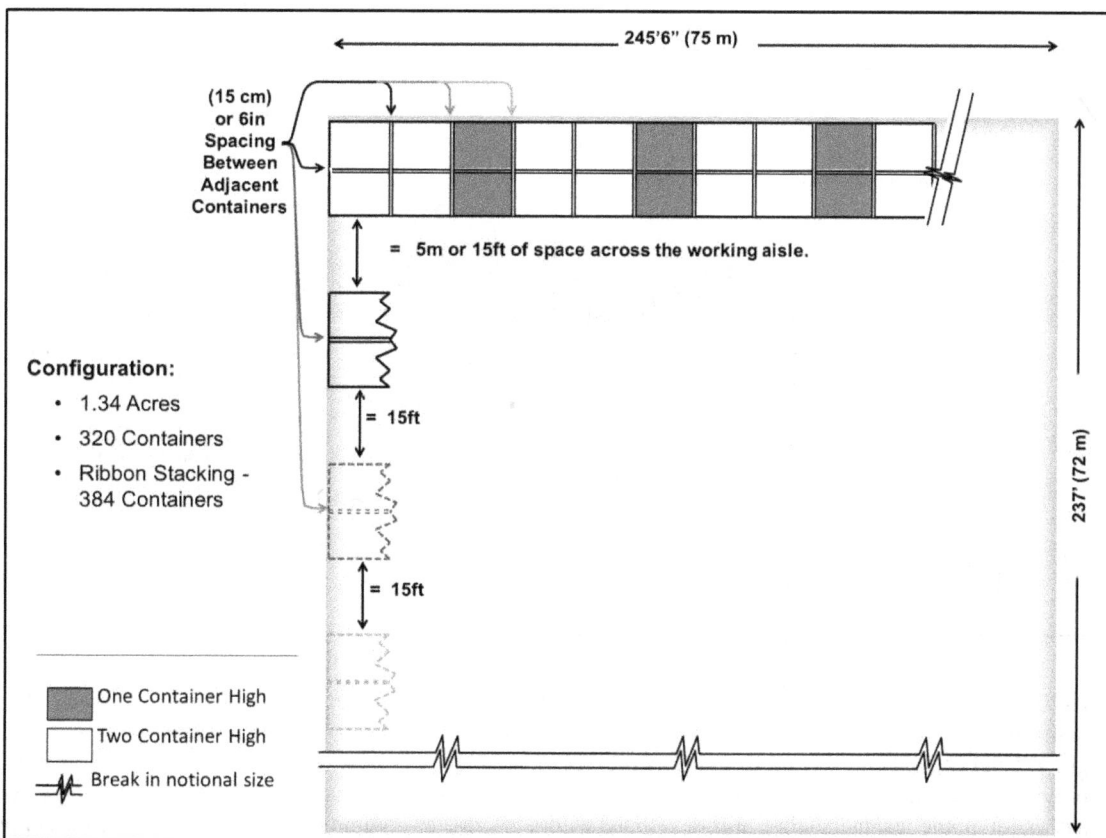

Figure A-9. Cluster plan for side loader turret stacking of 20-foot containers (15-foot working aisles)

Configuration:
- 1.34 Acres
- 320 Containers
- Ribbon Stacking - 384 Containers

245'6" (75 m)

(15 cm) or 6in Spacing Between Adjacent Containers

= 5m or 15ft of space across the working aisle.

= 15ft

= 15ft

237" (72 m)

- One Container High
- Two Container High
- Break in notional size

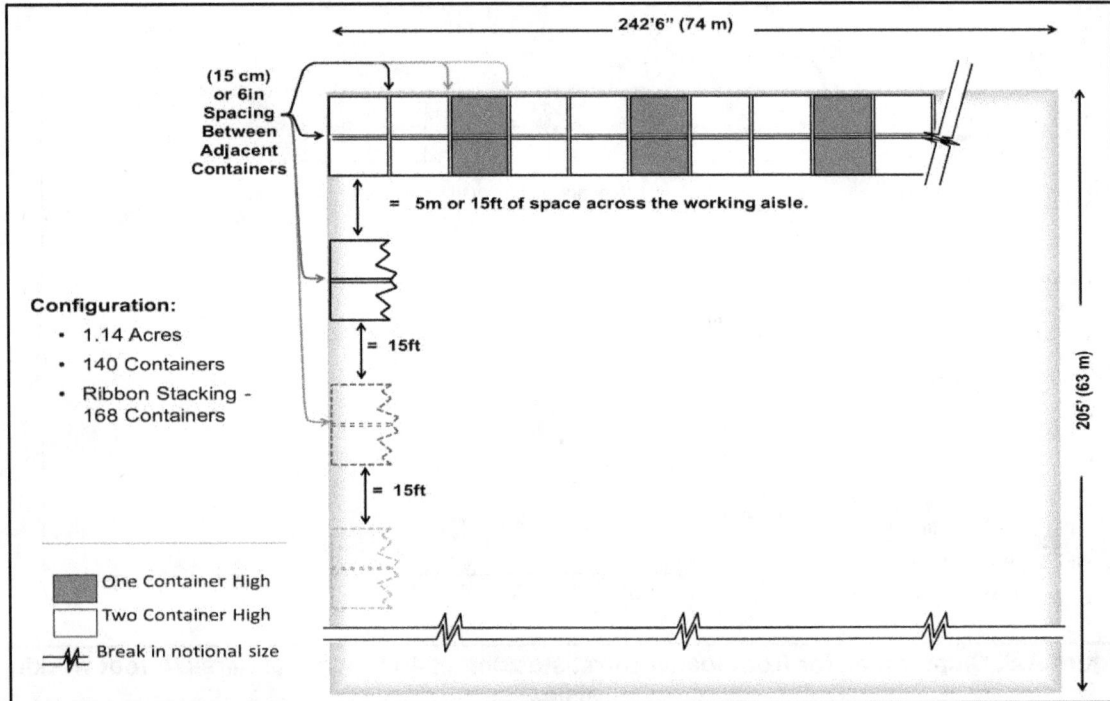

Figure A-10. Cluster plan for side loader turret stacking of 40-foot containers (15-foot working aisles)

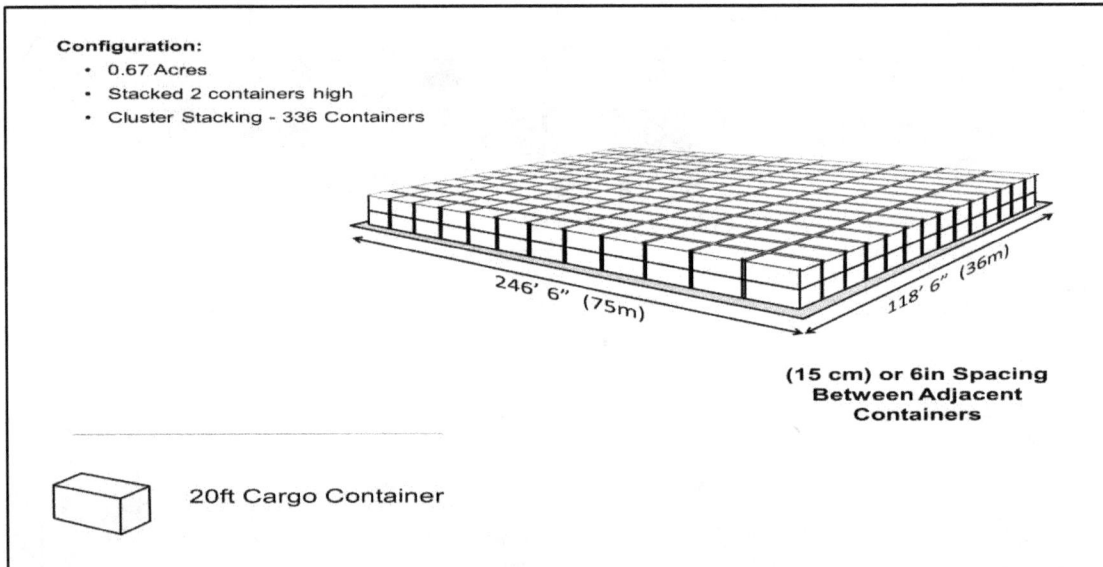

Figure A-11. 20-foot container cluster

A-20. Using the container cluster concept provides a relatively uncomplicated means of developing a marshaling yard commensurate with the needs of a specific operation or environment. This is done by grouping clusters within available space, modifying cluster dimensions where necessary, and adding areas to provide related activities. Figure A-12 shows a traffic pattern in an on-chassis marshaling area. Figure A-13 shows a hypothetical marshaling area developed within the cluster grouping concept. It is designed to support simultaneous discharge or back-load of one container ship in a permanent marine terminal operation. Intersecting aisles of the required width are placed around each separate container cluster. When two clusters are adjacent, they use a common intersecting aisle of the required width.

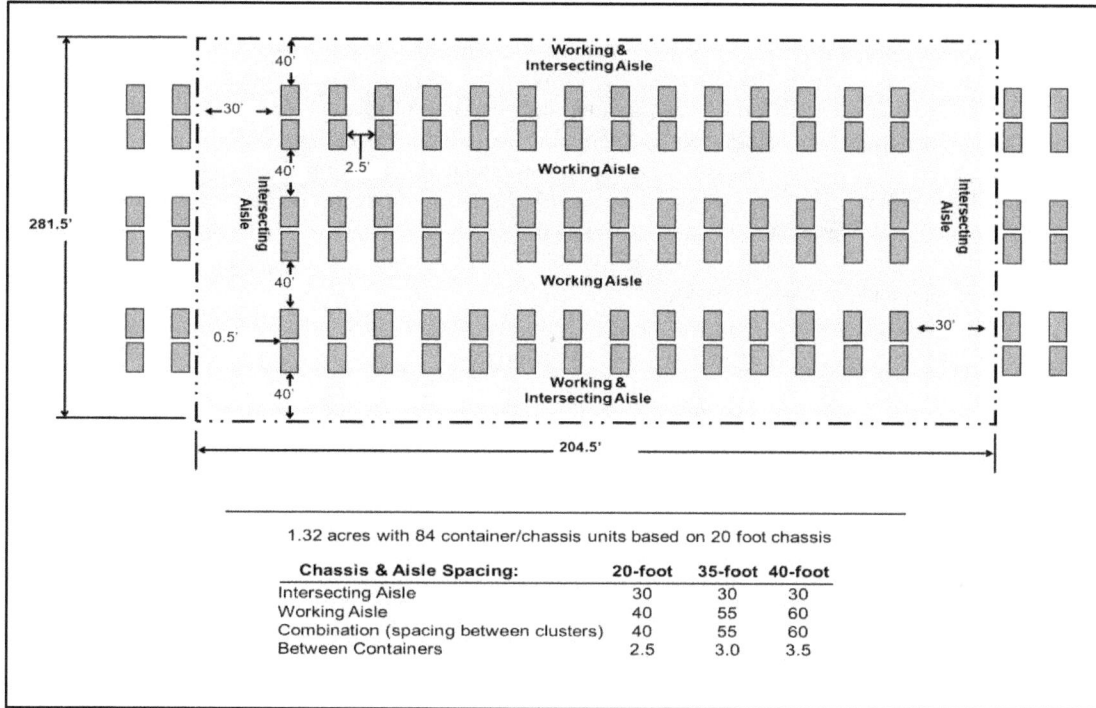

1.32 acres with 84 container/chassis units based on 20 foot chassis

Chassis & Aisle Spacing:	20-foot	35-foot	40-foot
Intersecting Aisle	30	30	30
Working Aisle	40	55	60
Combination (spacing between clusters)	40	55	60
Between Containers	2.5	3.0	3.5

Figure A-12. Traffic pattern in on-chassis marshaling area

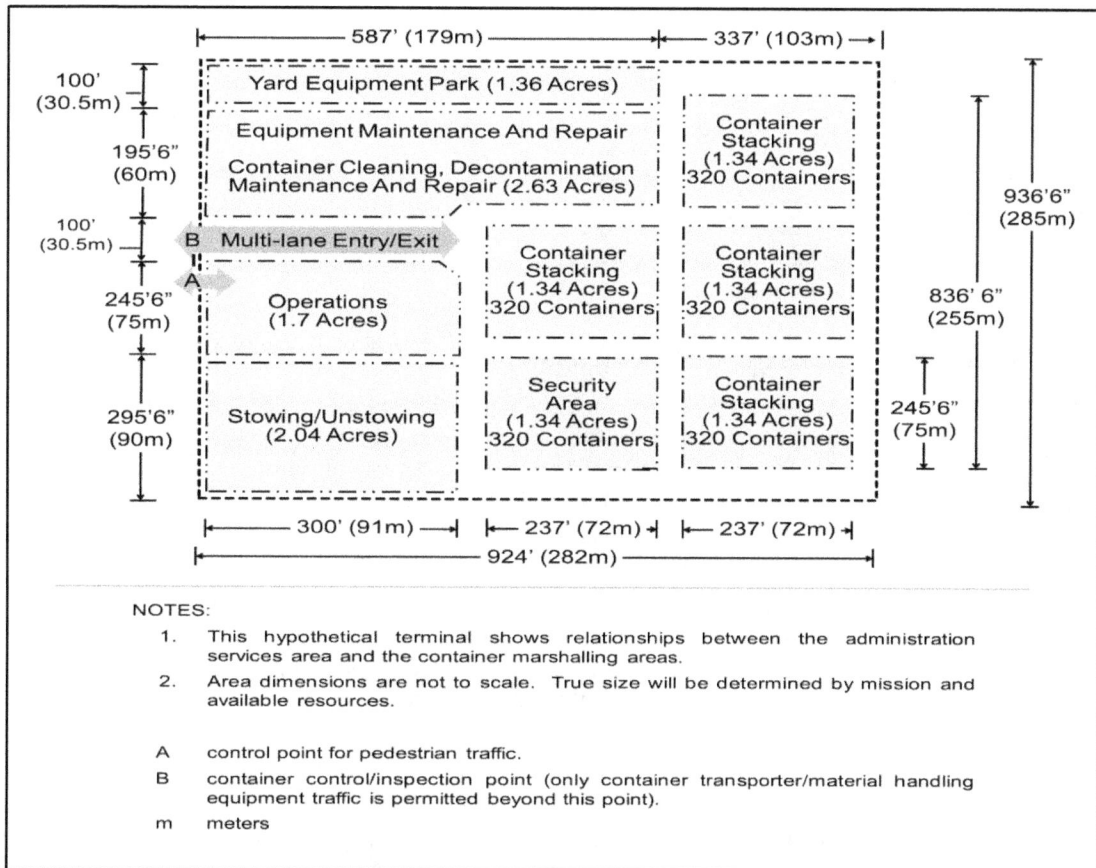

NOTES:

1. This hypothetical terminal shows relationships between the administration services area and the container marshalling areas.

2. Area dimensions are not to scale. True size will be determined by mission and available resources.

A   control point for pedestrian traffic.

B   container control/inspection point (only container transporter/material handling equipment traffic is permitted beyond this point).

m   meters

Figure A-13. Notional marshaling area

TERMINAL MARSHALING

A-21. The objective in any ship discharge terminal is to minimize the turnaround time of the ship. One way to do this is by always having the terminal tractors available and positioned properly at the cranes working the ship. To do this efficiently and with minimum congestion, the tractors should travel the least distance possible. The stacking areas should be located directly behind the crane's current working position at shipside. Hence, the two-deep stacking area can accommodate boxes from either crane as they work their way amidship. Each stacking area should be divided for import and export containers. Areas are divided to ease the drop-off of import containers and the pickup of export containers in one counterclockwise trip around the stacking area. One SOC works each container ship. Operating on a 24-hour basis, the unit should load or unload 600 containers per 24-hour period.

CONTAINER OFF-LOAD OR BACK-LOAD OPERATIONS

A-22. To off-load or back-load a containership, a minimum of two cranes will work each end of the ship in a coordinated effort. Each crane follows these steps in sequence for each hatch:
- Discharges all the containers on the hatch covers.
- Removes hatch covers.
- Discharges all containers from one cell.
- While discharging the next cell, back-loads the empty cell at the same time.
- Repeats all of the previous steps until all cells of that hatch are completed.
- Replaces hatch covers.
- Back-loads containers on hatch covers.

A-23. The SOC must maintain records of the stacking areas. These records give the specific location of each container within the terminal. A predetermined storage slot must be known for each container prior to it coming off of the ship. The actual space (numbers of clusters) required per ship berth depends mostly on the average dwell time of containers in the terminal. Potential bottlenecks in a marine terminal include—
- The dwell time for containers.
- Frustrated containers.
- Processing of containers at entrances and exits.
- Stuffing and unstuffing of containers.
- Cleaning and maintenance of containers.
- Method for container accountability.
- Vehicle delay and congestion.

MARSHALING AREA CLEARANCE OPERATIONS

A-24. Marshaling area clearance operations ensure containers flow rapidly and uniformly between dockside and the adjacent operational area and infrastructure. To minimize terminal congestion and work stoppages, marshaling area clearance operations are tailored to port unload or back-load output. An inbound container should not remain in the marshaling area longer than 24 hours. This also holds true for retrograde containers, provided a containership is available for back-loading. The normal procedure in clearance operations is to designate specific medium truck units to support a specific container unload or back-load operation.

A-25. The following paragraphs discuss motor transport requirements for marshaling area clearance support of one terminal service operation. Medium truck units as well as other truck companies operate around the clock (two shifts) with 75 percent equipment availability. The SOC unloads and, at the same time, back-loads 300 containers per day (two 10-hour shifts). Ideally, inbound containers should be cleared within 24 hours. If this is the case, a minimum of 300 containers per day must be cleared from the marshaling area. (For planning purposes, it is assumed that for each container moved from the marshaling area a retrograde container is returned.)

A-26. The traffic patterns within the terminal must be designed to support the cranes servicing a ship (see Figure A-14). Traffic patterns should be counterclockwise: up one side of the cluster when dropping off a container and down the other side when picking up a container.

Note: In Figure A-14:
1. Traffic patterns are designed to offer one-way traffic where possible to minimize distance and to eliminate congestion.
2. Area dimensions are not to scale.

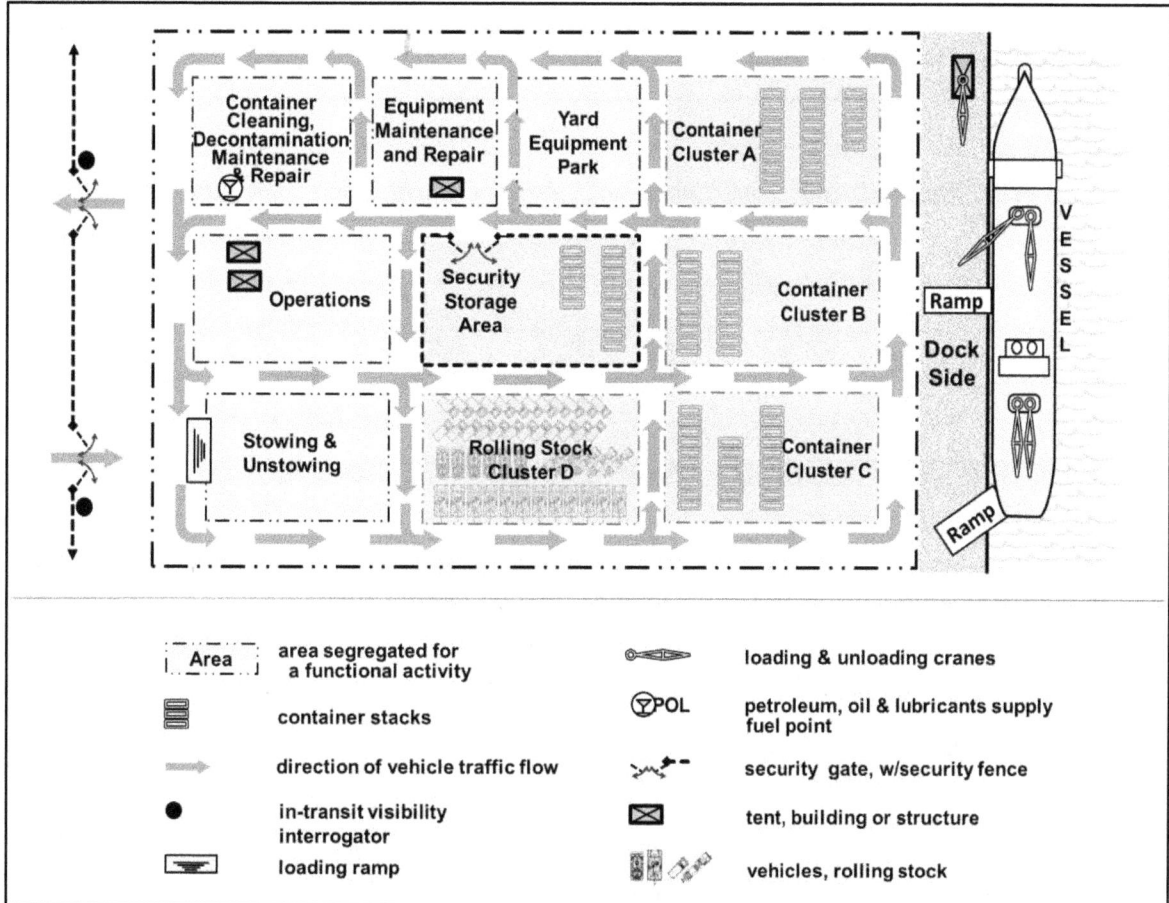

Figure A-14. Suggest traffic flow in permanent terminal marshaling area

RELATED SUPPORT

A-27. A transportation terminal battalion headquarters and headquarters company provides the basic operating headquarters for theater terminal operations. It is the normal command element for each two to four-ship marine terminal.

A-28. If it is a two-ship operation, a terminal battalion would operate the terminal. The battalion operations officer supervises consolidated battalion operations for documentation, inventory, and control functions. The battalion also controls operations of areas such as stowing, inspection, maintenance and repair activities, cleaning and decontamination, equipment parking, and security at battalion level. Thus, the terminal service companies can devote their efforts to container handling.

MARSHALING FOR RAIL MOVEMENT

A-29. Container movement by rail is used wherever possible. Rail presents a mass movement capability with little interference from weather or refugee traffic. Except for inland waterway, rail is the most economical mode for moving Army containers.

A-30. Figure A-15 depicts a procedure for marshaling, loading, and unloading containers for rail movement when the rail facilities are not a part of or adjacent to the marshaling yard. In the figure, retrograde containers are being exchanged for loaded containers, which will be moved inland. Before flatcars arrive, truck-transporters move loaded containers trackside, where they are pre-stacked two high as shown. After the flatcars are positioned, loading or unloading proceeds as follows: Container 1 moves to position B-1. Container A-1 is loaded in position 1. Container 2 moves to position B-2. Container A-2 is loaded in position 2. The process is repeated for the next flatcar. As the procedure continues, road C is used to remove retrograde containers to the marshalling area. This system is only used in certain circumstances When other equipment or circumstances prevail, other configurations are used.

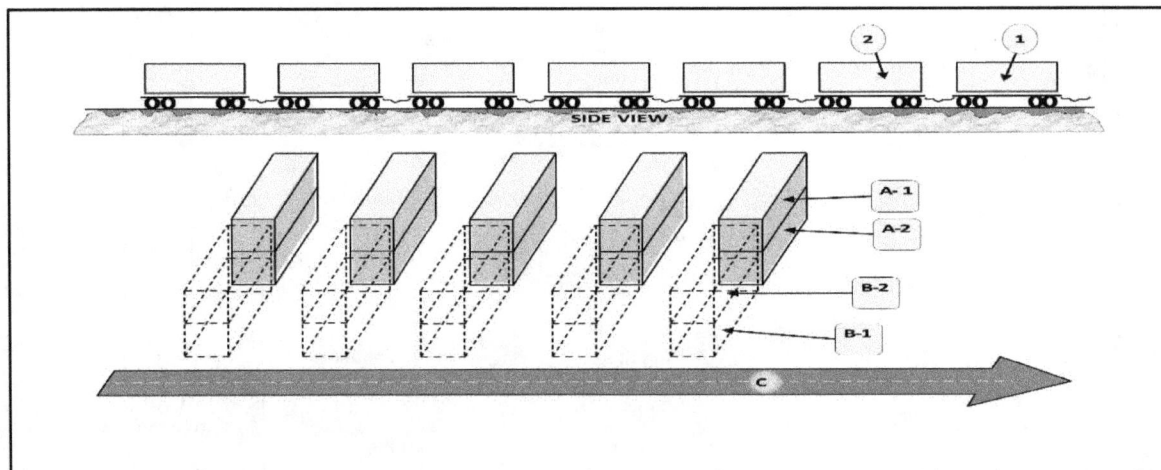

Figure A-15. Procedures for marshaling, loading, and unloading containers for rail movements when rail facilities are not part of or adjacent to the marshaling yard

DOCUMENTATION PROCESS

A-31. The marshaling yard documentation process is a vital link to create accurate automated documentation or accurately account for equipment if automated data processing is unavailable.

IMPORT CARGO

A-32. For import cargo, the shipping port transmits an advance manifest to the receiving port via the Global Air Transportation Execution System (formerly Worldwide Port System). Hatch summaries, preprinted from the advance manifest, provide the operator with advance notice of the types of cargo by size and quantity of incoming containers, movement priorities, and ultimate destinations. This information permits the operations officer to preplan marshaling yard space requirements and predetermine where each off-loaded container will be stacked. This is particularly important in planning the onward movement of outsize and overweight cargo.

STACKING LOCATION

A-33. Since the stack location of the container is planned, the cargo checker can receive a printout for the containers tallied. Using this as containers are unloaded from the ship, the cargo checker can direct the yard transporter to the designated stacking area. Radio communication between the cargo checker and the marshaling yard is the only way to ensure adequate control of the operation, especially in a large yard or in a highly flooded situation. If computer equipment is not available, a visual display board of the stacking area is kept by operations to provide container identification and location. A manual system requires appropriate internal communications.

CARGO DISPOSITION INSTRUCTIONS

A-34. These are used as a consignee advance notification document. Based on the cargo disposition instructions, the port's servicing MCT coordinates with the MCB through the ESC/TSC to arrange delivery dates and transportation to move containers from the marshaling area to final destination.

RETROGRADE MOVEMENT

A-35. When a retrograde container enters the marshaling yard, the container transporter driver presents the TCMD at the entry point and has the container inspected. The driver gets a receipted copy of the TCMD (proof of the delivery) and is directed to the point where the container is to be unloaded. The driver also gets a TCMD for the container that will be picked up for movement out of the yard. A TCMD is required each time cargo is moved from the area of responsibility. No container can be moved out of the marshaling yard exit or entry point without proper documentation and inspection. The container, the container transporter, and the container seal number must all agree with those shown on the TCMD. If not, the container will not be moved until proper documentation is prepared. When the container departs the marshaling yard, a copy of the TCMD is retained for entry into the central processing unit. It must be retained to show that the container has been shipped to the consignee and to update the computerized marshaling yard inventory.

SECURITY

A-36. Reduction of cargo theft and pilferage is a significant benefit of containerization. Compared to losses suffered in breakbulk operations, the reduction is indeed noteworthy. Security cargo should be stored separately from other cargo and should have its own secured area. Whenever possible, security cargo should also be unloaded from the ship during daylight hours. If possible, security personnel should observe unloading operations.

A-37. Figure A-16 on page A-16 provides a suggested design for security storage in a container marshaling yard. Ideally, the entire terminal should be enclosed. However, the security area should be enclosed at a minimum, preferably with a cyclone fence topped with several strands of barbed wire (concertina wire may be used as an expedient). If circumstances permit, a double fence should be installed. A 24-hour military guard should be placed on the gate. The perimeter should be patrolled periodically. Door-to-door placement of containers further strengthens security measures. Sensors, protective lighting, high-security locks, and access control procedures should be considered to help secure high-priority cargo. Adequate lighting and a sophisticated and constantly changing pass system greatly enhance security operations.

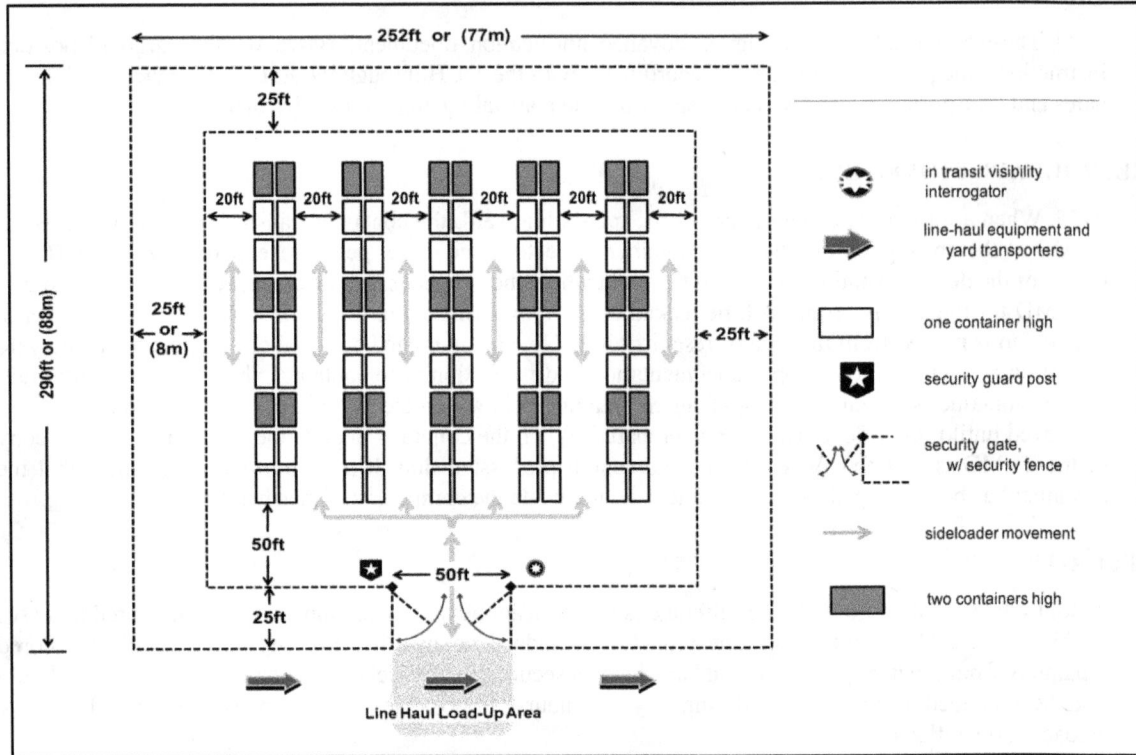

Figure A-16. Suggested design for a security storage area

Inbound and Outbound Traffic Control

A-38. Strict control of incoming and outgoing traffic is a key component of marshalling yard security. It is essential to restrict vehicular traffic entering or exiting the container stacking area to container transport equipment, MHE, and mobile scanning equipment. It is also essential to establish a single control point (gate) for vehicular traffic entering or exiting the container stacking area. This control point will be manned and operated by U.S. military personnel assisted by foreign national police or interpreters as necessary. U.S. military personnel (assisted by foreign national police or interpreters) should also operate a separate gate for pedestrian traffic. Surveillance and control functions of the vehicular control point include preventing entry of unauthorized vehicles and inspecting inbound and outbound containers. This is a thorough physical inspection including container condition; presence and condition of container seal or lock; evidence of illegal entry (such as tampering with or removal of door hinges); and, particularly for outbound containers, stolen items (look on top of and under the container and inspect the transporter cab). Other considerations include—

- **Documentation.** Was documentation checked for correctness, completeness, and legibility? (Ensure that transporter, container, and container seal numbers match those shown on the TCMD.)
- **Scanning equipment**. Were scanners used to read military shipping labels? (If there is no scanning capability, container numbers are reported manually to operations so that the yard inventory may be updated.)
- **ITV**. Were inbound and outbound cargo and equipment reported to the integrated data environment/global transportation network convergence?
- **Departure Time and Date.** Were the departure time and date for outbound containers entered on the TCMD and a copy retained for terminal files?
- **TCMD.** Was a signed copy of the inbound container TCMD provided to transporter operator to keep as a delivery receipt?

A-39. Surveillance and control functions of the pedestrian control point include the following:

- Permitting only authorized personnel to enter the container marshaling area.

- Maintaining, controlling, and safeguarding the pass system for foreign national personnel authorized to be in the area.

Perimeter

A-40. Security of the marshaling yard perimeter backs up gate security in keeping unauthorized persons out. Unauthorized persons may engage in sabotage, petty theft, and large-scale theft operations and may establish inside contacts with foreign nationals or other persons working in the yard. While it may not be possible to fence the entire yard perimeter, the security (sensitive, classified, or high-dollar-value cargo) area should be fenced with its own military-guarded gate. Perimeter defense measures may include one or a combination of the following:

- Chain-type fencing topped by three strands of barbed wire. (Inspect fence daily to ensure there are no holes or breaks.)
- Concertina wire.
- Sensors, when feasible.
- Patrols.

Container Transportation Operator

A-41. Drivers of the line-haul and local-haul container transporters are required to remain in the cab of their truck when operating within the container stacking area.

TRANSPORTATION CONTROL MOVEMENT DOCUMENT

A-42. No containers are allowed to move through the marshaling yard entry or exit control point without a valid and legible TCMD. Coordinating procedures prevent removal (either accidentally or purposely) of containerized cargo from the yard. At the gate, the container number is verified against the information provided by the movement and the documentation sections. The container, seal, and transporter numbers are verified for agreement with those entered on the TCMD. The container's seal is examined for breakage or evidence of tampering. Finally, the container is inspected for damage before it is released. A TCMD must also accompany retrograde containers. After control point personnel verify TCMD entries (such as container and seal numbers) and inspect the container, they give the driver a receipted copy of the TCMD. They also give directions to where the container is to be unloaded.

TCMD SECURITY

A-43. Blank TCMDs are not normally accountable by serial number. However, local procedures may serially number blank TCMDs as an added theft prevention measure. Regardless, blank TCMDs should be secured with one individual responsible for safeguarding and issuing them.

VERIFICATION OF CARGO ARRIVAL

A-44. Upon receipt of the container, the consignee returns a copy of the TCMD to the shipping terminal activity. The TCMD contains the consignee signature, date of receipt, and condition of cargo, container, and seal.

CONTAINER SEALS

A-45. A container seal is a device applied to the container door fastening. It indicates whether the door has been opened or the fastening tampered with, and if so, at what point in the movement system it happened. Seals are serially numbered to help identify the person who applied the seal and to provide a means of control. Failing to strictly account for seals from receipt to application defeats their purpose (to pinpoint unauthorized entry into containers). The following procedures promote container seal control and accountability:

- Maintain a record, by serial number, of seals received by the port operations officer and issued to authorized personnel for applying to containers.

- Store seals under lock. Designate one person to be responsible for the safekeeping, issuing, and recordkeeping of seals applied at the port.
- Designate specific persons (keep the number to a minimum) on each shift to apply seals and enter the serial number of the seal on the TCMD.
- Conduct periodic inventory of seals. Apply seals as soon as the container has been stuffed and as soon as a loaded (unsealed or improperly sealed) container is detected.
- Supervise the seal application. Failure to supervise or allowing a yard hustler to move an unsealed container to the stacking area offers an opportunity for pilferage of cargo before seals are applied. It also affords the opportunity to apply a bogus seal, break the seal later, remove cargo, and then apply a legitimate seal.

COMPUTING CONTAINER SPACE REQUIREMENTS

A-46. The following is a sample problem for computing container space requirements in a marshaling area:

A unit has been tasked to operate a container terminal with a total marshaling area of 830 feet wide and 886 feet long. The area must be designed for a one-ship operation using the side loader in the stacking clusters. To satisfy operational requirements, the stacking method must be used to enhance selective extraction. You are to determine the intrinsic capacity of the marshaling area using figures A-17 and A-18 (located on page A-20). Also use the same figures to perform the following steps:

- Step 1. Layout a plan of the area.
  - Draw a rectangle representing the area.
  - Draw in surrounding intersecting aisles.
  - Draw in through intersecting aisles.
  - Determine the measurements of clusters.
- Step 2. Determine the number of 20-foot containers in each row.
  - Determine how many 20-foot containers will fit into each row by dividing 340 (feet) by 20.5 (.5 equals half foot space allowed between containers for working room). This equals 16.58 containers per row. Any fraction is not counted a container; therefore, .58 is lost space (.58 x 20.5 = 11.89 feet). To provide more aisle space, move containers 10 feet to the left or right.
  - Stack containers (turret stacking) in two-two-one-high sequence in any given row. Every three ground slots have a five-container capacity. To determine the number of containers in a row, divide the number of columns by 3. Multiply that product by 5. If 3 does not divide evenly into the number of columns, the remainder is multiplied by 2 and added to the previous product.

    For example:

    16 columns divided by 3 = 5 (with a remainder of 1)

    5 x 5 = 25

    1 (remainder) x 2 = 2

    25 + 2 = intrinsic capacity of 27 TEUs per row in areas A and B

    Add the 10 feet of unused space to areas C and D
  - Repeat the calculation set forth in the previous paragraphs. For example:

    350 (feet) divided by 20.5 = 17.073 containers per row

    0.073 x 20.5 = 1.5 feet

    17 divided by 3 = 5 (with a remainder of 2)

    5 x 5 = 25 containers

    2 (remainder) x 2 = 4

    25 + 4 = 29 TEUs per row with 1.5 feet of unused space in areas C and D
- Step 3. Determine the number of rows.
  - Stacking 8-foot-wide containers side by side in double rows with a rolling space of .5 feet between the rows would occupy 16.5 feet. The side loader requires a 15-foot working aisle. So, in every 30.5 feet are stacked two rows. The length of this area is 368 feet, divided by

30.5 feet equals 11.65 or 11 double rows, with 21 feet remaining between a working aisle and an intersecting aisle.
- Using the intersecting aisle to work from would allow 16.5 feet of the 21 to be used for a further double row, for a total of 12 double rows.

Note: Each double row in A and B has 64 TEUs. Each double row in C and D has 68 TEUs. A and B each contain 64 TEUs multiplied by 12 double rows. This equals 646 TEUs in each quadrant. A and B together contain 1,296 TEUs. C and D each contain 68 TEUs multiplied by 12 rows. This equals 696 TEUs in each quadrant. C and D together contain 1,392 TEUs. A and B (1,296 TEUs) plus C and D (1,392 TEUs) equals and intrinsic capacity of 2,688 TEUs. The optimum operating capacity is 66 percent of 2,688 or 1,478 TEUs.

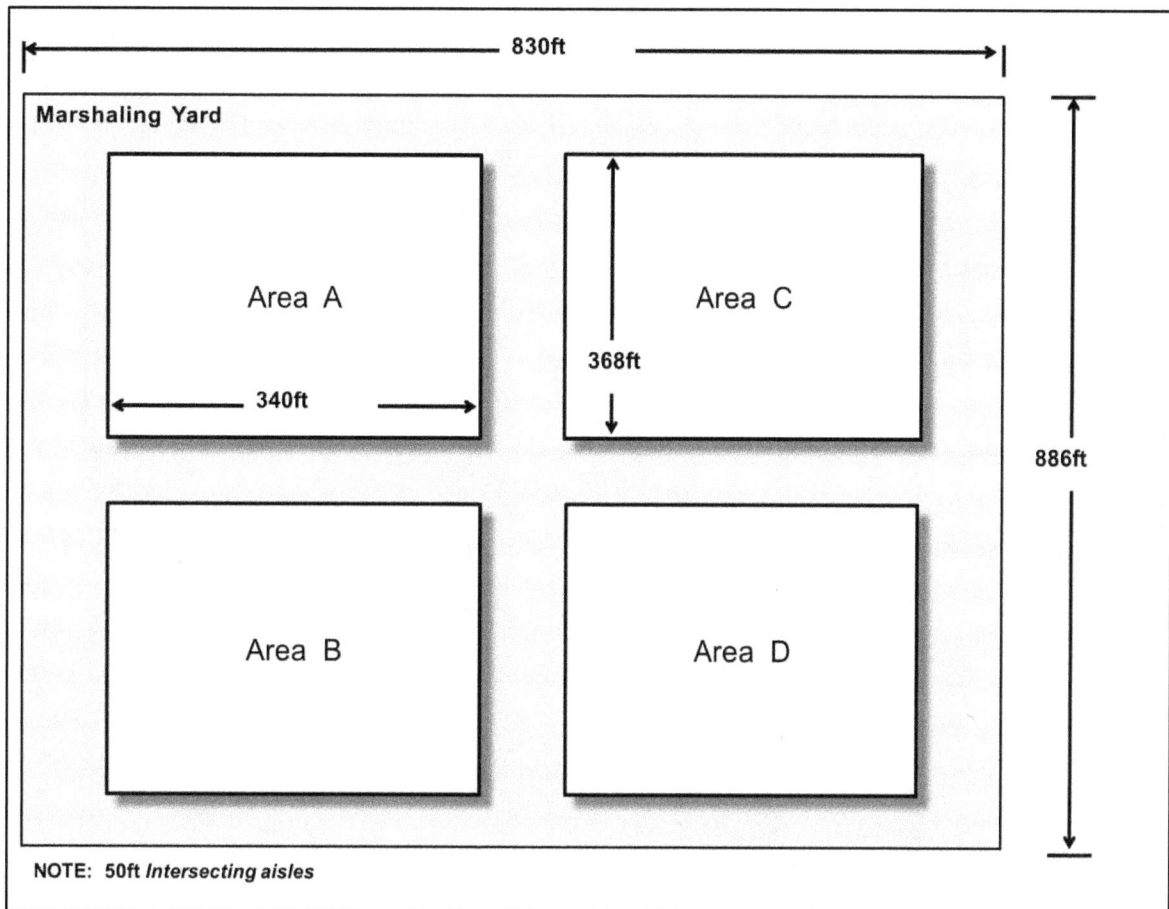

Figure A-17. Sample layout plan for container space requirements in a marshaling yard

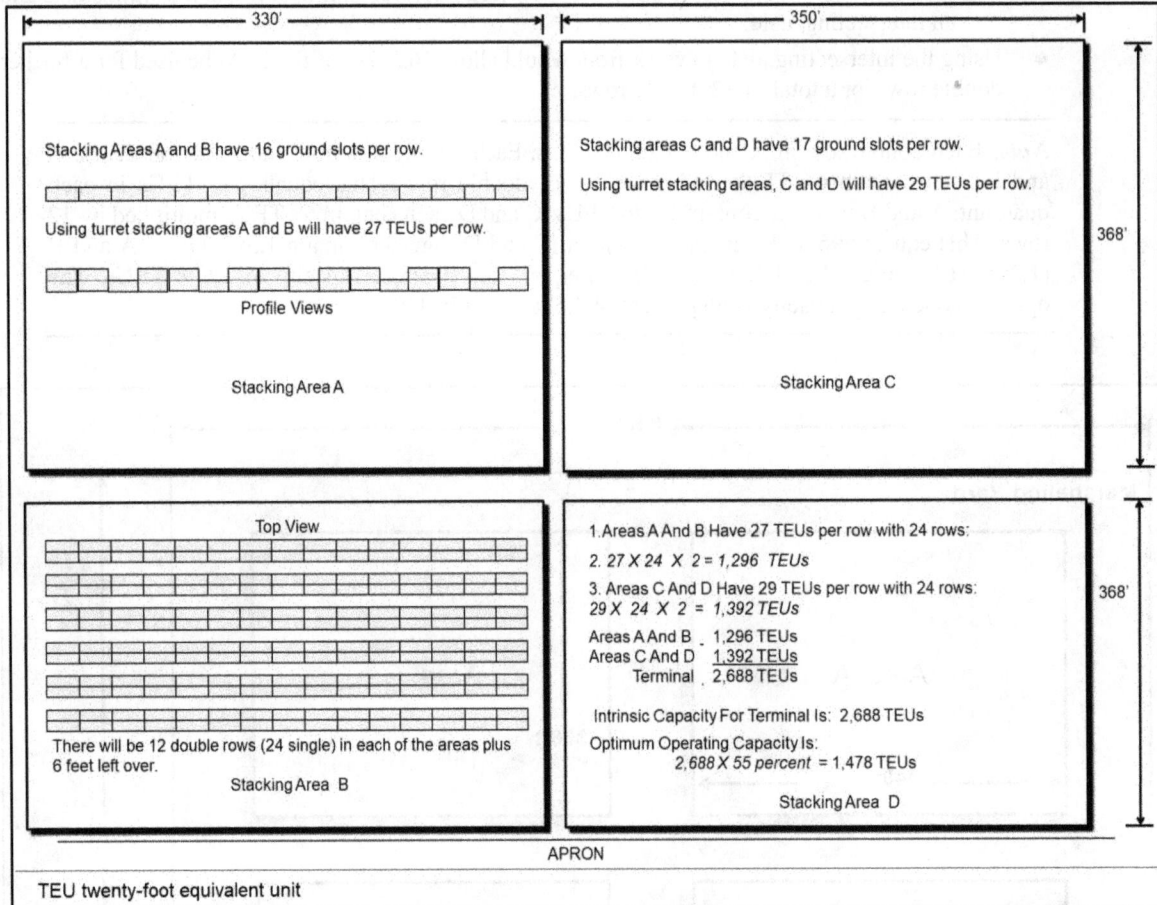

Figure A-18. Partial layout plan for container space requirements in a marshaling yard

Appendix B
Terminal Capacity

This appendix discusses terminal characteristics and capacities and how they are essential to determining the capability of a terminal to support operations.

THROUGHPUT CAPACITY

B-1. Terminal throughput capacity is critical in terms of determining the capability of a terminal to support operations. Estimating terminal throughput capacity is key to the process of selecting terminal sites and operating units. Terminal throughput capacity is the average quantity of cargo and passengers that can pass through a terminal daily—from arrival, to loading on the subsequent mode (for example, ship, air, truck, rail, bus, pipeline), to exiting the terminal complex. Factors, such as reception, discharge, transfer, storage, and clearance limitations may affect final throughput capacity. These are described in the following paragraphs.

RECEPTION

B-2. This capacity is a based on the number of ships (by type, length, and draft) that can be berthed in a harbor or at a terminal. Data used for estimating reception capacity includes—

- Channel depth.
- Channel width.
- Length of berths.
- Type of berths.
- Anchorage diameter.
- Water depth at berth.
- Type of berth.

B-3. Incoming ships are directed to specified terminals for discharge based on the type of ship, workloads or capacity of the terminal, and type and size of berths available.

- **Permanent Berths.** The type of ship selected to use permanent berths depends on the type of terminal at the berth; for example, container, breakbulk, or RO/RO. Vessels require 75 to 100 linear feet of berth length in addition to their measured length overall. Therefore, the longest vessel or combination of vessels must be 75 to 100 feet less than the length of the berth. The minimum water depth alongside the berth at mean low tide determines the maximum allowable draft for vessels at that berth. A ship should always have at least 2 feet under its keel for safety of the vessel.
- **Anchorage Berths.** For military planning, ships anchor either offshore or in-the-stream (harbor). Other methods exist, but these two are used for military planning purposes so the ship can get underway quickly. Use the formulas in Table B-1 to determine the required size (diameter) anchorage site for a ship.

Table B-1. Formula to determine diameter of anchorage site for a ship

Offshore: $D = 2(7d + L)$		In-the-stream: $D = R(4d + 2L)$
Use the following formula to determine the largest ship that will fit in a given area:		
Offshore: $L = (D - 7d)/2$		In-the-stream: $L = (D - 4d)/R/2$
D – diameter d – depth of water	L – length of ship R – reserve factor (1.1)	

DISCHARGE

B-4.   Discharge capacity is the cumulative amount of cargo that can be discharged from each terminal berth. This capacity is based on the capability of discharge methods and equipment used, as shown in Table B-2. Historical reports, shipper's reports, and realistic evaluations help in the estimation. The shortage of personnel must also be considered.

Table B-2. Terminal discharge capacity

Terminal/Berth	Discharge Capacity	Time Period	Method/Equipment
Breakbulk	2500 STONS breakbulk	24hrs or 1 day	CHE
Lighter berth	300 STONS breakbulk 450 STONS ammunition 200 containers	24hrs or 1 day	Army crane (1)
RO/RO berth	600 metric tons of cargo	3898 square feet per hour	Drive on/drive off
Underdeveloped container berth	300 containers	24hrs or 1 day	Heavy lift cranes
Developed fixed container terminal	700-800 containers	24hrs or 1 day	CHE (number varies by port)
CHE container handling equipment RO/RO roll on/roll off STONS short tons			

TRANSFER

B-5.   This is an evaluation of the capacity to move cargo from the discharge point to the storage point. It can be a time, equipment, and motion study that considers the number of moves available. For example, transfer capacity is the time it takes to move a pallet of cargo from the ship's side to the storage area, deposit it, and return to the ship's side. It is measured the same as the discharge capacity. When discharging ships at LOTS sites or anchored in the stream, transfer capacity is used twice, once for the lighterage and once with the MHE on the beach.

STORAGE

B-6.   This is the amount that can be stored at any one time. Storage capacity is given as an intrinsic capacity to obtain the operating capacity. The operating capacity depends greatly on the average dwell time of the cargo. Some cargo space must be left empty so that space is available to move cargo. Experience shows that congestion in the storage area begins at about 60 percent and is complete at 80 percent of the intrinsic cargo capacity of the terminal.

CLEARANCE

B-7.   This is the ability, measured like discharge capacity except by mode, to clear cargo from the terminal. Clearance conveyances for military purposes include, but are not limited to, trucks, railcars, lighters, and helicopters. The terminal clearance capacity may be limited by either of the following:

- Number of clearance conveyances.
- Ability of terminal equipment and personnel to load clearance conveyances.

VESSEL UNLOADING AND LOADING

B-8.   Based on the vessel manifest and cargo disposition instructions, the terminal battalion plans the discharge of individual ships before their arrival. This includes specifying the berthing location within the terminal and method of discharge. The terminal battalion works closely with the MCTs to ensure that variations from the vessel discharge plan are coordinated with mode operators to prevent delays in port clearance.

B-9. The harbormaster coordinates berthing, tug assistance, and employment of equipment and cranes required to discharge vessels.

B-10. Many factors affect unloading and loading of vessels. The sum of these positive and negative influences results in the actual capacity that can be achieved. These factors include—

- Weather.
- Sea state conditions.
- Visibility (fog and darkness).
- Crew experience.
- Lifting gear capability.
- Cargo stowage tactical situation.
- Terminal congestion.

B-11. The terminal commander's responsibility for outbound cargo is essentially the same for inbound cargo. The main difference is that the operation is performed in the reverse order. Loading a vessel, as does unloading one, includes for example—

- Coordination with shipping and transportation elements.
- Booking.
- Receiving.
- Stow planning.
- Cargo manifests.
- Preparing necessary shipping documentation.

B-12. Loading and unloading vessels is a complex operation that requires early planning and sustained communications between and throughout all elements in the chain of responsibility in order to execute a successful mission.

This page intentionally left blank.

Glossary

This glossary lists acronyms and terms with Army or joint definitions. Where Army and joint definitions differ, (Army) precedes the definition. The glossary lists terms for which ATP 4-13 is the proponent with an asterisk (*) before the term. For other terms, it lists the proponent publication in parentheses after the definition.

SECTION I – ACRONYMS AND ABBREVIATIONS

A/DACG	arrival/departure airfield control group
ADP	Army doctrine publication
AMC	Air Mobility Command
AO	area of operations
APOD	aerial port of debarkation
APOE	aerial port of embarkation
AR	Army regulation
ASCC	Army Service component command
ATP	Army techniques publication
CCDR	combatant commander
CCMD	combatant command
CFR	Code of Federal Regulations
CHE	container handling equipment
CRSP	centralized receiving and shipping point
CSSB	combat sustainment support battalion
DDSB	deployment and distribution support battalion
DDST	deployment and distribution support team
DOD	Department of Defense
DTR	Defense Transportation Regulation
ERC	expeditionary railway center
ESC	expeditionary sustainment command
FM	field manual
HN	host nation
IAW	in accordance with
ICTC	inland cargo transfer company
ITV	in-transit visibility
IWWS	inland waterway system
IWWT	inland waterway terminal
JDDOC	joint deployment and distribution operations center
JP	joint publication
JTF-PO	joint task force port opening

LOC	line of communications
LOTS	logistics over-the-shore
MCB	movement control battalion
MCT	movement control team
MHE	materials handling equipment
MSC	Military Sealift Command
OCONUS	outside the continental United States
ORP	ocean reception point
POD	port of debarkation
RFID	radio frequency identification
RO/RO	roll-on/roll-off
RSOI	reception, staging, onward movement, and integration
SDDC	Military Surface Deployment and Distribution Command
SOC	seaport operations company
SPOD	seaport of debarkation
SPOE	seaport of embarkation
STON	short ton
TBX	transportation brigade expeditionary
TCMD	transportation control and movement document
TDA	Table of Distribution and Allowance
TEU	twenty-foot equivalent unit
TM	technical manual
TSC	theater sustainment command
TTP	trailer transfer point
U.S.	United States
USAMC	United States Army Materiel Command
USAR	United States Army Reserve
USTRANSCOM	United States Transportation Command

SECTION II – TERMS

aerial port

An airfield that has been designated for the sustained air movement of personnel and materiel, as well as an authorized port for entrance into or departure from the country where located. (JP 3-36)

air terminal

A facility on an airfield that functions as an air transportation hub and accommodates the loading and unloading of airlift aircraft and the in-transit processing of traffic. (JP 3-36)

*beach capacity

The per day estimate expressed in terms of measurement tons, weight tons, or cargo unloaded over a designated strip of shore.

container management

(Army) The process of establishing and maintaining visibility and accountability of all cargo containers moving within the Defense Transportation System. (ADP 4-0)

Defense Transportation System

That portion of the worldwide transportation infrastructure that supports Department of Defense transportation needs. (JP 4-01)

intermodal

Type of international freight system that permits transshipping among sea, highway, rail, and air modes of transportation through use of American National Standards Institute and International Organization for Standardization containers, line-haul assets, and handling equipment. (JP 4-09)

intermodal operations

The process of using multiple modes (air, sea, highway, rail,) and conveyances (truck, barge, containers, pallets) to move troops, supplies and equipment through expeditionary entry points and the network of specialized transportation nodes to sustain land forces. (ADP 4-0)

joint deployment and distribution enterprise

The complex of equipment, procedures, doctrine, leaders, technical connectivity, information, shared knowledge, organizations, facilities, training, and materiel necessary to conduct joint distribution operations. (JP 4-0)

mode operations

The execution of movements using various conveyances (truck, lighterage, railcar, aircraft) to transport cargo. (ADP 4-0)

movement control

(Army) The dual process of committing allocated transportation assets and regulating movements according to command priorities to synchronize the distribution flow over lines of communications to sustain land forces. (ADP 4-0)

*multimodal

The movement of cargo and personnel using two or more transportation methods (air, highway, rail, sea) from point of origin to destination.

port of debarkation

The geographic point at which cargo or personnel are discharged. (JP 4-0)

port of embarkation

The geographic point in a routing scheme from which cargo or personnel depart. (JP 3-36)

port opening

The ability to establish, initially operate and facilitate throughput for ports of debarkation to support unified land operations. (ADP 4-0)

staging

Assembling, holding, and organizing arriving personnel, equipment, and sustaining materiel in preparation for onward movement. (JP 3-35)

staging area

A general locality established for the concentration of troop units and transient personnel between movements over the lines of communications. (JP 3-35)

theater opening

The ability to establish and operate ports of debarkation (air, sea, and rail), to establish a distribution system and sustainment bases, and to facilitate throughput for the reception, staging, and onward movement of forces within a theater of operations. (ADP 4-0)

This page intentionally left blank.

References

All websites accessed 27 March 2023.

REQUIRED PUBLICATIONS

These documents must be available to intended users of this publication.

Department of Defense Dictionary of Military and Associated Terms. May 2023.

FM 1-02.1. *Operational Terms.* 09 March 2021.

FM 1-02.2. *Military Symbols.* 18 May 2022.

RELATED PUBLICATIONS

These documents are cited in this publication.

JOINT PUBLICATIONS

Most Department of Defense publications are available online: https://www.esd.whs.mil/DD/.
Most joint publications are available online: http://www.jcs.mil/Doctrine.

DODI 4715.22. *Environmental Management Policy for Contingency Locations.* 18 February 2016.

DODM 4715.05, Volume 1. *Overseas Environmental Baseline Guidance Document: Conservation.* 29 June 2020.

JP 3-31. *Joint Land Operations.* 3 October 2019.

JP 3-34. *Joint Engineer Operations.* 6 January 2016.

JP 3-35. *Joint Deployment and Redeployment Operations.* 31 March 2022.

JP 3-36. *Joint Air Mobility and Sealift Operations.* 4 January 2021.

JP 4-0. *Joint Logistics.* 4 February 2019.

JP 4-01. *The Defense Transportation System.* 18 July 2017.

JP 4-09. *Distribution Operations.* 14 March 2019.

JP 4-10. *Operational Contract Support.* 4 March 2019.

ARMY PUBLICATIONS

Most Army doctrinal publications are available online: https://armypubs.army.mil/.

ADP 3-0. *Operations.* 31 July 2019.

ADP 4-0. *Sustainment.* 31 July 2019.

AR 200-1. *Environmental Protection and Enhancement.* 13 December 2007.

ATP 3-34.5/ MCRP 3-40B.2. *Environmental Considerations.* 10 August 2015.

ATP 3-35. *Army Deployment and Redeployment.* 9 March 2023.

ATP 3-93. *Theater Army Operations.* 27 August 2021.

ATP 4-02.1. *Army Medical Logistics.* 29 October 2015.

ATP 4-10/ MCRP 4-11H/ NTTP 4-09.1/ AFTTP 3-2.41. *Multi-service Tactics, Techniques, and Procedures for Operational Contract Support.* 16 December 2021.

ATP 4-11. *Army Motor Transport Operations.* 14 August 2020.

ATP 4-12. *Army Container Operations.* 12 February 2021.

ATP 4-14. *Expeditionary Railway Center Operations.* 22 June 2022.

ATP 4-15. *Army Watercraft Operations.* 3 April 2015.

ATP 4-16. *Movement Control.* 25 April 2022.

ATP 4-43. *Petroleum Supply Operations.* 18 March 2022.

ATP 4-48. *Aerial Delivery.* 21 December 2016.

FM 3-34. *Engineer Operations.* 18 December 2020.

FM 3-94. *Armies, Corps, and Division Operations.* 23 July 2021.

FM 4-0. *Sustainment Operations.* 31 July 2019.

FM 6-27/MCTP 11-10C. *The Commander's Handbook on the Law of Land Warfare.* 07 August 2019.

TM 3-34.56/ MCRP 3-40B.7. *Waste Management for Deployed Forces.* 29 March 2019.

OTHER PUBLICATIONS

Codes of Federal Regulations are available online: https://www.ecfr.gov/.

32 CFR, Part 651. *Environmental Analysis of Army Actions (AR 200-2).* 23 January 2023.
https://www.ecfr.gov/current/title-32/subtitle-A/chapter-V/subchapter-K

Defense Transportation Regulations are available online: https://www.ustranscom.mil/dtr/

DTR 4500.9-R-Part II. *Cargo Movement.* May 2014.

DTR 4500.9-R-Part V. *Department of Defense Customs and Border Clearance Policies and Procedures.* June 2021.

PRESCRIBED FORMS

This section contains no entries.

REFERENCED FORMS

Unless otherwise indicated, DA forms are available on the Army Publishing Directorate (APD) website (https://armypubs.army.mil); DD forms are available on the Executive Services Directorate (ESD), Washington Headquarters Services (WHS) website (https://www.esd.whs.mil/directives/forms/).

DA Form 2028. *Recommended Changes to Publications and Blank Forms.*

DD Form 1384. *Transportation Control and Movement Document.*

Index

Entries are by paragraph number.

A-H

air terminals,
 A/DACG, 4-14–4-20
 operations, 4-2–4-4

C

containers,
 operations, A-10
 stacking configurations, A-11

I-K

intermodal operations,
 components, 1-7
 planning considerations, 1-31
 port opening, 1-36
 terminals, 1-23

L

land terminal operations,

executing, 3-4
 planning, 3-2
land terminal types, 3-5–3-32
logistics over-the-shore
 operations,
 planning, 6-2
 operations, 6-20

M-N

maritime terminals,
 planning, 5-12
 operations, 5-12–5-15
 types, 5-2–5-7
marshaling yard,
 location, A-6
 operations, A1–A3
 security, A-36
 TCMD, A-42

O-S

organizations, 2-6–2-32

T-U

throughput capacity, B-1
transportation brigade
 expeditionary,
 capabilities, 7-5
 concept of operations, 7-2
 mission command, 7-3

V

vessel unloading and loading, B-8

W-Z

water port opening, 5-8
water terminal operations, 5-16

This page intentionally left blank.

By Order of the Secretary of the Army:

JAMES C. MCCONVILLE
General, United States Army
Chief of Staff

Official:

MARK F. AVERILL
Administrative Assistant
to the Secretary of the Army
2316411

DISTRIBUTION:
Active Army, Army National Guard, and United States Army Reserve. To be distributed in accordance with the initial distribution number (IDN) 345344, requirements for ATP 4-13.

This page intentionally left blank.

This page intentionally left blank.

This page intentionally left blank.

PIN: 104103-000